JUST ONE RAIN AWAY

Just One Rain Away

The Ethnography of River-City Flood Control

STEPHANIE C. KANE

McGill-Queen's University Press
Montreal & Kingston • London • Chicago

© McGill-Queen's University Press 2022

ISBN 978-0-2280-1427-0 (cloth)
ISBN 978-0-2280-1428-7 (paper)
ISBN 978-0-2280-1529-1 (ePDF)
ISBN 978-0-2280-1530-7 (ePUB)

Legal deposit fourth quarter 2022
Bibliothèque nationale du Québec

Printed in Canada on acid-free paper that is 100% ancient forest free (100% post-consumer recycled), processed chlorine free

Funded by the Government of Canada / Financé par le gouvernement du Canada  Canada Council for the Arts / Conseil des arts du Canada

We acknowledge the support of the Canada Council for the Arts.

Nous remercions le Conseil des arts du Canada de son soutien.

Library and Archives Canada Cataloguing in Publication

Title: Just one rain away : the ethnography of river city flood control / Stephanie C. Kane
Names: Kane, Stephanie C., author.
Description: Includes bibliographical references and index.
Identifiers: Canadiana (print) 20220284814 | Canadiana (ebook) 20220284911 | ISBN 9780228014270 (cloth) | ISBN 9780228014287 (paper) | ISBN 9780228015291 (ePDF) | ISBN 9780228015307 (ePUB)
Subjects: LCSH: Floods—Manitoba—Winnipeg. | LCSH: Rivers—Social aspects—Manitoba—Winnipeg. | LCSH: Flood control—Manitoba—Winnipeg. | LCSH: Flood control—Social aspects—Manitoba—Winnipeg. | LCSH: Flood control—Manitoba—Winnipeg—Decision making.
Classification: LCC GB1399.5.C3 K36 2022 | DDC 551.48/909712743—dc23

This book was typeset by True to Type in 10.5/13 Sabon

To C. Jason Dotson, with love

Contents

Table and Figures ix
Preface xi
Acknowledgments xv

1 Geo-Culture in Winnipeg, Manitoba 3
2 Ice/Water on the Move (Prehistory) 45
3 Systems and Seasons (1950 and 1997 Floods) 77
4 Falling out of Synch (1997 Flood) 101
5 Furies of Wind and Wave (2011 Flood) 115
6 Sunlight Machines in Floating Ecologies (Diatoms) 142
7 River-Trickster on TV (2014 Flood) 156
8 Avulsion! 181

APPENDICES
I Archive and Anthropocene Dialogue 199
II Technozones 207
III Press Conference Transcript 211

Notes 221
References 281
Index 303

Table and Figures

TABLE

7.1 Temporal jurisdictions within clock time 162

FIGURES

1.1 The Flood Bowl. Drawing by author based on map elements from "The Forks Riverbank Restoration Project," an outdoor sign by Manitoba Parks 4
1.2 Pelicans at Portage Diversion Spillway, Assiniboine River. Photo by author 8
1.3 Riverbank stabilization, Lyndale Drive, Red River. Photo by author 14
1.4 Wall Through Time, The Forks. Photo by author 24
1.5 The actual confluence, The Forks. Photo by author 40
2.1 Double artifact: Geoscience (A) + port city projection (B). Composite image drawn by author. Geovisual data from (A) Teller, Leverington, and Mann (2002, 880) and (B) CentrePort, https://centreportcanada.ca/about-centreport/ 66
2.2 Erratic boulder, Portage Avenue. Photo by author 73
3.1 Provincial flood control system. Digital image (made on PowerPoint) by author, adapted from Li and Simonovic (2002, 2,655) ad MIT (2013b, 11) 80
3.2 Hydrometric river network. Drawing by author based on Debi Forlanski's impromptu interview sketch, 8 October 2014. Photograph by author 96

5.1 Cottage three years after the Flood of 2011, south shore of Lake Manitoba. Photo by author 120
7.1 Chronotope of encounter (emergency press conference). Drawing by author 166
8.1 Assiniboine avulsions. Digital image (made on PowerPoint) by author adapted from Rannie et al. (1989) and MIT (2013b, 18) 185
AII.1 Intersecting technozones. Drawing by author with hydrology base map from Weir, ed. (1983, 17) 209

Preface

As a cultural anthropologist of water, I've walked the river edges of Colón, Veracruz, Amsterdam, Santos, Salvador, Buenos Aires, Kochi, Valparaíso, Singapore, Pisco, Zagreb, and the place at the centre of this book – Winnipeg, Manitoba.

Governing the water flowing through all these cities are three basic infrastructural systems – potable water, sewage, and drainage. The systems seem everywhere the same, more or less, each designed to fit, each constructed and operated through an alliance of engineering and law based on geoscience. Before Winnipeg, I would take the fact of this basic sameness as a logistical boon. The tripartite infrastructures functioned like a ready-made script. Against their sameness, I could plot difference; against their leaden materiality, I could put the symbolic into flight. As I went from place to place sketching out grids, I would slip between back stages and front stages. I'd focus in-depth participant observation, the essential ethnographic method, in culturally significant and politically active neighbourhood waterfronts and also in the agencies that governed their waters. Fieldwork would unfold, project by project, shared river-city stories would begin speaking to each other in an alchemy of crossing scales, languages, and cultural ecologies.

But Winnipeg was a surprise, a puzzle within a puzzle within a puzzle. It's amazing that there is such a vibrant, sophisticated, and arty city sitting in a flood bowl on the northern edge of urbanization. I focused on Winnipeg's state-of-the art flood control system that, like in river cities everywhere, has yet to be redesigned and rebuilt for the intensifying, more frequent, and unpredictable floods already accompanying the climate crisis. Floods present the ultimate engineering quandary: naturally occurring throughout history yet occurring more and more *un*naturally now. How to reinvent the city to prepare for climate change in the face of such uncertainty?

Before Winnipeg, most of the other cities I studied were sited *where rivers meet the sea* (the title of my last book), so flood management was always complicated by tidal surges. Winnipeg, in the continental centre of North America, has no tidal surges. But it does have *ice*. Enter ice/water as a lively ethnographic character. Ice confounds calculations of all sorts. *Ice jams* in spring will block north-flowing currents and force the water out of the channel onto the prairie. When the river abandons its body to become a sea, all bets are off until it retreats back into its historical channel. I imagine these ice/water encounters as spring re-enactments of ice age end, when retreating glaciers blocked their own meltwaters from rushing to the sea. Ice/water thus pulls the imagination back into geological time and forward into unpredictable futures. Ice/water cultivates an ethnographic sense of cities and rivers as geological actors entwined in what I call the sphere of unintended agencies.

Thriving despite its icy spring floods, Winnipeg fulfills expectations yet provokes independence of thought. The city forces one to think again about whatever it was that one had in mind. Winnipeg's quiet but sure provocation is hard to pin down (and for all I know, this may be a Canadian thing). In my case, Winnipeg changed my approach to ethnographic experimentation.

In a nutshell, instead of using technical knowledge as an empirical baseline for multi-city, global ethnography, I turned it into a cultural artifact for analysis. No longer would I ride side-by-side with the discursive power of global water governance and its technical dependencies. No longer would I work to fill their empty grids with social, cultural, and political description and analyses. The grid itself is a geo-cultural artifact, a thing, a model, a metaphor. Are there ways to loosen the grip of this grid and grid-like thinking, to decolonize the clearly unjust foundations of modern monumental engineering? Are there ways to work with geoscience, engineering, and law without relinquishing the felt desire and responsibility to relate to rivers as living beings and to respect the sovereignty of Indigenous territories? Here, on these pages, I do this earthwork in prose and image.

Pragmatic urban ecology meets experimental ethnography in this book. I have managed to poke some holes in the technical black boxes of flood control expertise. I am not a scientist, engineer, or lawyer;

Acknowledgments

Special thanks to the institutions that financially supported my residency as chair in Environmental Science at the University of Winnipeg's Richardson College for the Environment and Science in the fall of 2014: the Council for the International Exchange of Scholars (CIES) and Fulbright Canada and the College of Arts and Sciences and Office of the Vice Provost for Research, Indiana University Bloomington. Thanks, too, to the Department of International Studies and the Hamilton Lugar School of Global and International Studies for helping to fund publication of this book.

I am deeply grateful to the people and institutions that contributed to my ethnographic field research and enhanced my experience of everyday life in Winnipeg and the province of Manitoba. I am especially indebted to Danny Blair, Pauline Greenhill, and Stephen Kohm for being inspiring hosts and colleagues and for welcoming me into their social and professional networks. Thanks to the many persons, named within, who generously shared the invaluable expert-inhabitant knowledge that permeates this book. Thanks, too, to supportive organizations that lent substance to this work's core and enhanced my broader cultural, political, and ecological understanding: in addition to the academic Departments of Women's and Gender Studies, Criminal Justice, and Geography and Richardson College at the University of Winnipeg, they include the Ministry of Infrastructure and Transport (MIT, now called MTI), Naturalist Service Office, Water Survey of Canada, Ecowest/CDEM (Conseil de dévelop-pement économique des municipalités bilingues du Manitoba), Manitoba Hydro, Stantec, Public Interest Law Centre, The Forks, Oak Hammock Marsh Interpretive Centre, Living Prairie Museum, CentrePort, North Red Waterway

Maintenance, home of the Amphibex, Institute of Urban Studies Library, Local History Room in the Winnipeg Public Library, Manitoba Archives and the Legislative Library, Canadian Museum of Human Rights, the galleries hosting the lively exhibits and events of MAWA (Mentoring Artists for Women's Art), and the rural municipalities of Dunnattor, Morris, and St Adolphe. For their critical feedback on work-in-progress, I'd like to thank audience members at the talks I gave at the Prairie Division of the Canadian Geography Association in Riding Mountain National Park, at the University of Winnipeg, and in the Winnipeg Public Library's Skywalk Series.

In the years of analysis and writing between fieldwork's end and the finished book manuscript, my academic home, the Department of International Studies, chaired by Purnima Bose and Padraic Kenney in the Hamilton Lugar School of Global and International Studies, provided a steady stream of scholarly resources, encouragement, wonderful colleagues, and students. Beyond my home campus, this book project has benefitted from wide-ranging, interdisciplinary interactions. Participation in two key research networks dramatically influenced my geological imagination. The Ice Law Project opened my eyes to the plural interpretations of ice/water's geophysics and geopolitics (ILP, convened by the Centre for Border Research at Durham University, UK, with Leverhulme Trust funding, 2014–19). Special thanks to Phil Steinberg, its visionary director, and my talented co-leaders on the Mobilities and Migrations subproject: Claudio Aporta, Aldo Chircop, and Kate Coddington. The Rivers of the Anthropocene, a 2018 workshop/symposium at the IUPUI Arts and Humanities Institute, marked the beginning of amazing transdisciplinary conversations with Jaia Syvitski and, later, with Kimberly Rogers, both of whom seek to incorporate human dimensions into earth science models. My thinking about ice/water infrastructures and urban ecologies has benefitted from comments and questions emerging from meetings of the American Anthropological Association, American Association of Geographers, Community Surface Dynamics Modeling System of the University of Colorado Boulder, International Congress of Arctic Social Science, Nordic Geographers, and Croatian Ethnological Society. There were workshops and talks, too, including (listed chronologically from past to present): Department of Ethnology and Cultural Anthropology Lecture, University of Zagreb; Urban Waterscapes Workshop, University of California San Diego; conference on Global Issues-Cultural Perspectives, Willem however, I do believe that if the struggle to redesign the flood control system and its operations gathers force in Winnipeg or elsewhere, some of these ethnographic holes might provide specific starting points for reinventing terms that have been languishing inside basic urban technicalities like hydrological balance. More generally, this ethnography reframes human–water relationships and illuminates the planetary significance of river cities. Acknowledging the powerful and pervasive workings of the sphere of unintended agencies – where many of the crises of climate originate and reside – will help, I argue, to clear pathways to climate adaptation through climate justice. And so, there the book begins.

Stephanie C. Kane
Bloomington

Pompe Institute and the Institute of Criminal Law, Utrecht University; conference on Infrastructures as Regulation, Institute for International Law and Justice, NYU Law School; ILP workshop on Questioning Territory, SUNY Albany; Cultural Studies Common Seminar and Colloquium, University of Pittsburgh; conference and workshop on Hydrohumanities, University of North Texas; Sea Change, digital conference of the Society for the Anthropology of Consciousness; Un/Predictable Environments, digital conference hosted by University of British Columbia Vancouver, Queen's University Belfast, and the University of Allahabad. There were many opportunities to share work-in-progress within my department and across the IU Bloomington campus, e.g., Ambiguous Geographies, Center for the Study of Global Change; Unforeseen Constellations, the Polish Center; PhD Students' Conference, School for Public and Environmental Affairs; Being Human in the Age of Humans, Humanities without Walls; Rethinking Globalization, Media School Graduate Association. Each one of those events opened my eyes to perspectives that contributed to the shape of this book. Thank you.

Many thanks go out to the dedicated professionals at McGill-Queen's University Press who worked with diligence and thoughtful good cheer throughout the COVID-19 pandemic. I am especially grateful to acquisition editor Khadija Coxon and my two anonymous reviewers.

And to my family, born and made – the Kane-Nichols, Michler, Savage, Burbank-Cohen, Mason, and Dotson clans – thanks for all the love, food, joy, and beauty. Please keep it coming! And of course, thanks to my beloved husband and partner in all things, C. Jason Dotson.

Finally, I give thanks to the Indigenous peoples and their ancestors on both sides of the Canada-US border for being first and being present.

Parts of chapters in this book were previously published as parts of single-authored chapters:

2018. "Where Sheets of Water Intersect: Infrastructural Logistics and Sensibilities in Winnipeg, Manitoba."
2022. "Winnipeg's Aspirational Port and the Future of Arctic Shipping (The Geo-Cultural Version)."

JUST ONE RAIN AWAY

1

Geo-Culture in Winnipeg, Manitoba

I. INTRODUCTION

Flood events put everyday lives of inhabitants of river cities like Winnipeg, Manitoba, in tension with water's impulsive, implacable, elemental force. Yet whether in extreme disorder or in balance, cities and rivers share states-of-being; they are beings-in-tension. When rivers rise up out of their familiar channels to become freshwater seas, submerging landscapes of human habitation, then receding from neighbourhoods left forlorn, they enact riverhood. When again and again, cities reshape themselves around flood-prone rivers, they enact human collectivity. Impulses of rivers and human collectives (cities) combine as they carry on being and becoming, persisting in distinctive yet entwined embodied form.[1] In this sense, water bodies and collective humanimal bodies share in the planet's geophysical dynamics – which must also be *geocultural* dynamics – as they move and persevere together. Rivers braid into and out of cultures' interpretive frames (see figure 1.1).

Cities and rivers enact a common impulse to sustain their distinct yet flexible forms. In the process, human intentions meet unpredictable events. I name the dimension of matter and meaning within which cities and rivers enact common impulses and within which events often unfold unpredictably, the *sphere of unintended agencies*. Flood control, then, can be defined as an intentional collective act to keep the city dry. In other words, flood control is a collective human effort to invent and sustain an altered state-of-being shared with rivers.[2]

Knowledge from expert fields of geoscience, engineering, and law, together with inhabitant experience, provide ethnographic material for telling flood control stories. These convey a geo-cultural imaginary with human and more-than-human actors who decipher, democratize,

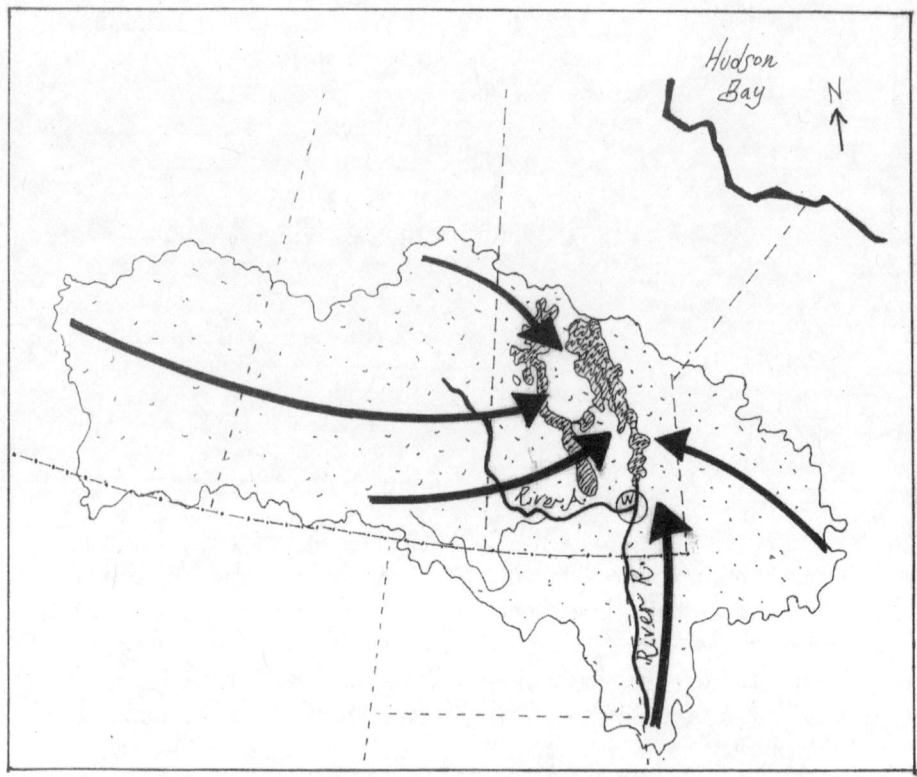

1.1 The Flood Bowl. Water flows into the continental interior from multiple directions (arrows) and eventually out to Hudson Bay. The city of Winnipeg sits inside the Lake Winnipeg watershed (dotted area) at the confluence of the Red River (ending in Lake Winnipeg) and the Assiniboine River, which flows eastward across the prairie toward the city past Lake Manitoba's south end (lakes = scribbled areas).

and reenchant technical expertise for a new kind of appreciation. For the fact is, while technical flood control knowledge and its spatial logics seem straightforward and practical, intentions and effects can be out of synch for elemental reasons not always clearly recognized. Throughout this book, the sphere of unintended agencies works as a touchstone, a reminder to appreciate the impressive capabilities of flood control alongside a critical analysis of its assumptions, limits, sacrifices, and colonial erasures. And too, the sphere of unintended agencies, where cities and rivers coexist in-tension but only humans act intentionally, provides an opening for attempting to tell socio-technonature stories of water bodies as if they were as alive as you and I.[3]

To begin, I briefly sketch two entryways into the geo-cultural imaginary: the first scene enacts glacial retreat and human appearance unfolding as the ice age ends; the second, a fractal, unfolds inside the expansive spacetime of the first to join the ethnographer as she first walks into a node within the Manitoban flood control system. Together, the two scenes stretch ethnography into co-occurring spacetimes from which glaciers, rivers, and cities emerge as actors on a geological stage that is both specific to Manitoba and to planet earth.[4] The first scene tells a geoscience origin story of the aquatic forces that sculpted the earth's surface in the configuration familiar to humans. In the midst of the transition between Holocene and Pleistocene Epochs, ice/water plays the lead. In the second late-summer scene, the ethnographer introduces herself to a tentacle of the city's flood control infrastructure. According to the official chronology of geological time, the second scene comes after the first by roughly ten millennia. The scenes' juxtaposition on the page is designed to cultivate a sense of both chronology and simultaneity. Deep time past – immanent in present experience and future projections – foreshadows the empirical work world of engineers and forecasters (see chapter 3).

A. *The Outburst Quartet*, an Origin Story

Not your normal ethnographic scene (#1) in which ice/water rivers of the northern hemisphere enact the clamouring flows of climatic past-present-futures.

> Only cracking ice and howling winds break the silence, or so I imagine ice age sounds with the help of geoscience. Up in the Arctic, the weight of practically infinite snow falls, flake on flake, layer on layer. Gravity and cold inexorably compress snow's soft crystalline mounds into solid ice. With increasing weight and thickness, a hard, frozen mass pushes the thinner ice in the front outward and down along the planet's contours: glaciers on the move (Upham 1895, 108). In the midst of Pleistocene's epochal cold, the enormous lobed glaciers combine into the Laurentian Ice Sheet to cover almost all North America. As they heave their nosey weight across the surface, the ice polishes, grinds, grabs, scratches, cutting striations into bedrock. Millennia upon millennia pass until the next major climate change when temperatures begin to rise, bringing the Holocene, the geoscience-mother

epoch of our species. Then, changing once again, the continental ice sheet starts to rot. As glaciers lose size and retreat north again, their icy lobes discharge cold meltwater from subglacial channels. Trapped, though, by the glaciers out of which they melt, the waters fill the northward-tilting trough in the continent's interior. The largest of the meltwater bodies formed, named Lake Agassiz, leaves strand lines in the sand and on the cliffs as it drowns, at one time or another, more than a million km^2 of terra firma.[5]

Eventually, in sudden and chaotic episodes, the ice walls confining Lake Agassiz give way. Lake Agassiz bursts in four directions with volumetric might. Awesome silence fills the pauses between sudden sounds of cracking and crashing of ice, rushing and gurgling of fast-moving currents: a four-movement expression of ice/water's massive phase-shifting struggle. The meltwater outbursts follow one after another in four different directions – Mackenzie Valley to the Arctic Ocean, Hudson Bay to the North Atlantic Ocean, St Lawrence to the North Atlantic Ocean, and the Mississippi Valley to the Gulf of Mexico – quite possibly changing into the arrangement of ocean currents that ships sail today. Enough of Lake Agassiz remains in the trough to (re)form what will become the most important water bodies in Winnipeg: Red and Assiniboine Rivers, Lakes Winnipeg and Manitoba.

Bison herds appear in this new hypothetical landscape born of glacial retreat. They stand watch as waves of a new bipedal mammalian species disperse into prairie wetland.[6] Some thousands of years later, descendants of bison herds and first peoples are met by waves of settlers bringing sedentary traditions from across the Atlantic Ocean. Unlike first peoples who move toward and away from the aquatic descendants of glaciers fluxing their volumetric powers, the settlers see disruptive rivers-in-flood and vie for mastery of prairie terrain. They build flourishing yet vulnerable river cities armoured with flood control infrastructure. Gentler river-city relations recede into history and hinterlands. Now long-dead bison herds haunt city streets, rematerialized as icons travelling solo in flat colour-coded corporate logos and gallery art.[7]

Whoosh! Whoosh! Whoosh! Whoosh! ... glacial meltwater tentacles crash through jagged walls of ice into four oceans, disrupting currents and raising levels. Wow. To commemorate the geo-cultural significance of Teller, Leverington, and Mann's (2002) hypothesis on the way to becoming scientific fact, its implications for understanding unintentional agencies of ice/water and, too, the potential of this multiphase event as grand metaphor, I name it the Outburst Quartet. The four-directional outbursts from continental centre to oceans follow one upon another and through time to reverberate in the land and seascapes of the present. Ice/water's form and cyclic phase-shifting potential is, I believe, an inherent liveliness reborn in every spring ice breakup and every major flood. I name this set of end-ice-age events the Outburst Quartet as a way to convey its mythical-scientific postglacial power and to recognize it as the founding impulse of flood control. This ten-thousand- (or so) year-old continental drama sets the stage for all the flood dramas since before time immemorial, that is, long before settlers and their written history, long before first peoples and their oral histories, before bodies had names. Outburst Quartet, as metaphor inspired by four outburst floods modelled by geoscientists at ice age end, reappears throughout the book simply as a useful portal into rough and tumble geological spacetimes wherein ice/water actors sometimes hold, sometimes relinquish, their powers to create and destroy humanity's expectations.[8]

B. *An Engineered Tableau*

An ethnographer begins to get the lay of the land in this not quite normal ethnographic scene (#2). Roughly following the Assiniboine River upstream on 27 August 2014, I drive west out of Winnipeg across the prairie along Trans-Canada Highway 1 (= Portage Avenue) to the Portage Diversion (see figure 1.2).

> In continuous, steady motion, water rushes through the giant concrete walls of the reservoir's spillway and falls into the churning riverbed below. A pelican squadron floats in a little side swirl of water between base and bank. Long orange bills plunge down, then up into the air, throat pouches full of deranged fish. Just downstream, a man fishes with a rod. Where line meets surface, a barely perceptible circle. On the grassy hill adjacent, a woman and small children play on a blanket.[9] A stranger-ethnographer stands apart. (Traveling from her home in Bloomington, Indiana, she is a NYC-born third-generation settler-descendant of immigrants fleeing from no-return Eurasian elsewheres.)[10] Her camera silently shoots the engineered tableau.

1.2 Pelicans fishing at the base of the Portage Diversion Spillway – an unintended ecological impact of Assiniboine River engineering.

I see now as I write: pelicans, man, family – my fellow spillway visitors – teach me an early lesson: monumental infrastructures of concrete and steel unintentionally offer sustenance and connection in ways that exceed engineers' intentions. For creating an excellent fishing spot is surely not an intentional effect of this multi-million-dollar node in Manitoba's flood control system.[11]

Inside the fieldwork scene but outside the city, I wonder about riverhood as im-pulsive entity and about the com-pulsive discipline of river + control structure (although I do not yet have these terms in mind). Because I have prepared for this site visit, a cognitive map of the whole system mentally supplements my ground view: the dam-spillway structure in the tableau is part of a larger assemblage called the Portage Diversion. A key infrastructural node in the provincial system, the Portage Diversion shunts floodwaters north to Lake Manitoba and away from Winnipeg to the east.[12] Stuck in this relationship, the Assiniboine seems to live a hemmed-in existence, divided and diminished all year long, even when not flooding. Yet the river's impulses have power too: they continue to organize and motivate the spatial distribution of pelicans, people, fish and, too, determine the control structure's siting, design, and operational routines.

An infrastructural question comes to mind: how are riverine impulses entangled with smaller, mobile animal and plant bodies within the scene? The river buoys up floating pelicans, parting its surface and splashing around them when they dive into the water between base and bank. Currents carry fish downstream to meet the hook a man knots to the end of a line: together, river (unintentionally) and man (intentionally) may trick fish into biting. Out of range of touch, family will presumably cook and eat caught fish, absorbing river water into their collective bloodstream. For her part, after contemplating the anthropogenic ground – dam set to allow river flow through spillway, earthen bank sculpted to fit around control structure, mowed grass – the ethnographer carries the river away as digital imprints, later to choose one view among others to share with readers, though because of financial constraints in book publishing, the photo's black and white rendering erases the vibrant contrast of bright orange pelican beaks against the sunlit river's muddy colors.[13]

But there is so much more going on in this habitat of myriad living beings. Among the unseen, consider diatoms, opalescent single-celled creatures invisible to the naked eye but accessible if a limnologist or botanist collects, preserves, and prepares water samples for microscopic evaluation. Diatoms are ubiquitous in freshwater systems, although their species diversity is so specific, they can be used in forensic identification of locations where criminal suspects deliver corpses to the deep.[14] The diatoms must have been there in the water and must also have been affected by the river + control structure. Imagine, then, diatoms floating in the reservoir's calm layers near the top, collectively using their glass-walled bodies to turn sunlight into fish food until, oh no! some stray too close to the spillway and tumble willy-nilly into the churning river channel, where, laden down by sediment, their work is interrupted – until eventually, currents carry them into Lake Manitoba where quieter waters allow them to renew their biological (but oh so magical) purpose.

I learn about diatoms only later, during the writing process (see chapter 6), but I invite them into this opening ethnographic scene because their presence, invisible but ecologically foundational, dips this little story into the sphere of unintended agencies where primary producers labour. For who really knows what the engineering effects of providing new lake-like spaces for diatoms to flourish and for pelicans to gorge on fish – at first glance, positive effects – might do to ecological sustainability of lacustrine and fluvial habitat? Also, not within my consciousness while in scene #2 is scene #1. Writer-me knows that scene #2 is a human-scale fractal of scene #1's geological spacetime: the afterlife of the Outburst Quartet moves within and without this assemblage of river, control structure, pelicans, people, fishing rod, camera, diatoms. How else to appreciate the Assiniboine's unintended agency holistically than my method – juxtaposition of increasingly unlikely subjects and ways of knowing?

There is one more twist, a geological fact that links urban river language to deep time: the Assiniboine flowed through "here" before the Outburst Quartet's tentacular performance. The Assiniboine has flowed through this space, this internal piece of continental crust, since before the last ice age and its requiem, before any humans imbued this space with meanings that came to signify a place. Afterwards, when a much grander, more powerful but younger Red River emerged from the bottom of Lake Agassiz, the Assiniboine slipped down the hydrogeological hierarchy to become a mere tributary. However, given its longevity, its long length, its trickster nature, I suggest that it should not be too surprising that the Assiniboine is the river to be acting queerly, the patterned impulse sustaining its familiar meandering noticeably different now. This river, unique in having lived through climate change past, is sending the clearest signals of climate change present. The notion that in some circumstances, a long but volumetrically smaller river may have a greater talent for issuing climatic communiques than a short, large one may be a hypothesis worth considering scientifically. But I do admit, at this point, I arrive at the end of this double beginning more through the logic of story than through detailed evidence from geoscience.

I define geo-culture as the combined logistical domains of professional and everyday life that persistently work to imagine and translate the technical into the territorial and fearful emotions into rational systems of control or influence. Geo-culture provides a framework that stretches understanding of contemporary cities and urbanization into deep time and considers expert-inhabitants as geological as well as social actors.[15] The -culture of geo-culture operates through frameworks of interpretation, including the techno-scientific frameworks at the core of flood control. Rather than allowing techno-science to remain apart in its own domain and supplementing it with meaning from social science and humanities domains, this ethnographer edges into and across its expert-inhabitant boundaries.[16] Insofar as I can find and navigate openings, I strive to decode the cultural power of techno-scientific knowledge and practice.

An invented genre of ethnographic exploration, geo-culture builds on traditions that interpret "material social processes of culture" as "ways of life" (Williams 1977, 19). Grounded in participant observation, interviews, and embodied landscape interpretation, the writing stretches experimentally by juxtaposing all kinds of texts, imagery, and models of ice/water and urban earthworks. I keep the dash in geo-culture to keep the conceptual space between them open. Like other scholars confronted with the wider implications of Anthropocene geoscience, though much more down to earth, I see geo-culture resonating with theoretical writings such as Yusoff's (2016, 6) "geosocial" gatherings of collective material subjectivity and Povinelli's (2016, 4–6) critique of the "geontologies" of life and nonlife as defined in late liberalism. Writing ethnography as geo-culture, I invite a wide range of readers into the subject of flood control, highlight the extraordinary accomplishments of Winnipeggers, demystify technical subjects, and engage the geological imagination.

This book explores known and unknown realms of geo-cultural imagination by unravelling the tactical and mass mediated alliances of geoscience, engineering, and law that bring flood control into being. I write in this conceptual terrain: I build on empirical evidence that geoscientists produce by narrowing, systematizing, and replicating knowledge within frames of reference that sometimes edge into the speculative (e.g., the Outburst Quartet). I trace the appearance and application of geoscience in flood control engineering, law, and everyday life. Then, exploring taken-for-granted ideas and practices basic to flood control, I pose possible alternatives. The ground of provincial authority, I argue, comes alive when analysis of empirical knowledge practices that support it include the mythologies of its making. Hence the need for stories.

II. GETTING TO KNOW THE RED RIVER VALLEY OF THE NORTH: MAPS, MEMORIES, AND EVERYDAY LIFE

Framing the ethnographic scenes to follow: Whenever I conceptualize a new project, I always begin with a visit to my university library map room with its orderly rows of grey metal cabinets diligently labelled by region and country. Piles of cartographic offerings in a jumble of time and style fill every drawer, waiting to be chosen, lifted out with care, and pored over. I love the feel of big paper maps, some made of cloth and finished for habitual handling, so thing-like compared to digital versions in the catalogue. I bring home a colourful selection and lay them about my study until the spacetimes they materialize seep into my consciousness, forming a preview of the local geo-culture I hope to discover. As a result of such preparations, a cognitive map accompanies me when I meet Winnipeg's rivers in person for the first time as ethnographic subjects. This cognitive map infiltrates the new maps I buy when I arrive to do fieldwork.

A funny thing happens on this trip, though. The actual river (it is along a bend in the Red that these events take place) rejects the cognitive model I bring along. This is not because the cartographers had made more than the usual mistakes or over-generalizations in their rendering of the river's linear relation to city streets. And I'm not referring to ground-proofing, wherein adjustments for accuracy depend on eliminating differences between aerial camera lens and human eye. Rather, the river simply would not cooperate with the fixity and angle of cartographic vision. The Red's stubborn unreadability triggers subtle confusion that, with some attentiveness, leads me to rediscover something about my maps and about probably all maps. They carry the potential for error-making when human bodies navigate actual, unfamiliar water bodies. The gap between map-based cognitive knowledge and knowledge gained through walking with a living river, occasionally conversing with neighbours, leads me into a sphere of unintended agencies within geo-culture. That said, most Winnipeggers would find the telling of everyday life and local history that follows quite familiar.

A. Neighbourhood Meander

I rent a furnished house a few paces from the Red River as a fieldwork base, arriving in the heat of August 2014 and leaving at the end of November after only a little taste of winter's cold. Like many of my neighbours and their dogs, I make a habit of walking along the riverbank promenade along Lyndale Drive, watching the water flow and flocks of birds turn in the sky. Although I travel throughout the city and province doing interviews and infrastructure site visits, I get to know this stretch of river just south of The Forks better than any other. I feel how the Red acts upon all the senses, how it organizes an outdoor aesthetic by sharing its open vistas and rippling conversations with clouds, canoes, and trees rooted in its banks. As all those who regularly walk the promenade know, under certain conditions, the river might rise up and overwhelm the congenial scene and the neighbourhood's well-kept middle-class homes nearby. In ordinary times, the river's aesthetic may recede into the background of this or that human drama, which, except for a couple of intense rainstorms, is the case during my stay. At other times, especially in early spring, the Red may act in concert with forecasters and news media to generate the anticipation of catastrophe.[17]

By personal experience or profession, the walkers inhabit their particular version of "metropolitan nature" (Gandy 2002). They call on their own theories of the invisible causes of the territorial instabilities with which they are most familiar (like the easily ignored but potentially ominous spread of cracks in the asphalt road beside the promenade that a neighbour points out to me one day).[18] When I explain that I have come to their city to study advances and conundrums in urban flooding and flood control, people laugh and would say something like: "Well, you've come to the right place!"

This scene of nature in the city draws dog-walkers out of their homes even in the dead of winter, as one woman says, "when the wind is cold enough to rip your face off." But though it is a compelling social space lending character to the neighbourhood, the promenade we walk upon is not a natural levee created by sediment-carrying current. It is in fact a primary dike, part of the twentieth-century provincial infrastructural system that protects the city from devastating floods. I did not properly understand these street-shapes in the landscape as water infrastructure until some days after arrival, when they are deciphered for me by Bill Rannie, a professor from the University of Winnipeg's Department of Geography and expert on regional hydrogeology.[19] Day by day, as I photograph the river's edge, I gradually became familiar with the prosaic arrangements of bleached logs, stone, cement, and metal stabilizing the more troublesome grass-covered sections of neighbourhood earthworks (see figure 1.3). Such material traces of structural maintenance silently narrate a history of the city's ongoing negotiation with the river passing through it. A few blocks walk inland, a baseball diamond sits in the depression made by the earth's removal, the dike's subtracted twin.

1.3 Stone, timber, metal, and plantings stabilize the Red River's clay bank along Lyndale Drive, a mundane but critical portion of the city's primary dike system.

Never having lived in a city organized by the confluence of two meandering rivers, I had to adjust my habitual ways of orienting in space. Looking across the stretch of river along the promenade, I see that my neighbourhood is separated from the neighbourhood on the other side, which creates a visual nearness but tactical distance. Without a canoe, getting to the other side entails a vehicle, roads, and a bridge. Orientation changes in deep winter too. When the river freezes hard enough to be traversed on foot, skates, sleds, and trucks, the ice brings perception and landscape into alignment and opens up possibilities for new relationships.[20] For example, one promenade walker told me how she found her fluffy white dog, which a few winters ago was quite lost after having wandered across the ice from an unidentifiable home on the other side.

Besides imposing a division that must be traversed, the river's meanders introduce cognitive complexity, triggering confusions of scale and directionality. For the first several weeks that I joined the walkers along the promenade, I couldn't figure out the pattern of ripples on the river's surface. I knew that the wind direction could be out of synch with the current, but even when there was no wind the river didn't seem to be moving from south to north as it is supposed to. One day, after asking a number of people why the ripples were radically opposite to the pattern one would expect, most merely shrugged. But one woman looked me in the eye and explained kindly, "You know, my dear, the river runs east-to-west on this stretch, yes? This mixes things up a bit." Ding!

Maps present us with a reality based on accumulated knowledge, but what we know from maps is different from what we see, hear, and feel by experiencing a place firsthand (Wood 2010, 16). In this early phase of getting to know the river, I had taken the maps I used at face value, improperly transposing what I thought I knew to what I sensed in situ. Thinking back to what led to the error, I believe it came from the map of Manitoba that presents the Red River in an international scale emphasizing the vertical axis. Crossing the forty-ninth parallel from the Dakotas into Canada at Emerson, the Red flows north through Winnipeg along Provincial Trunk Highway 75 (a.k.a. Lord Selkirk Highway) to its end in the world's eleventh-largest freshwater lake, Lake Winnipeg. This flat, paper map influenced the cognitive map I carried around in my head when I first walked the promenade. It led me to transpose the simplified vertically aligned image from the international scale on to the neighbourhood's east-west meander.[21]

The error in transposing scale, however, is not simply due to lack of familiarity or expertise. Nor is it due to the way road maps simplify rivers. As I would learn in interviews, technological advances that render topographic data more precise can make it even more difficult to sustain accuracy while transposing scales.[22] A major challenge in forecasting extreme flood events, for example, involves following moving targets like river crests while changing scales. As a flood approaches, digital data indicating river stage and flow intensity stream in via satellite to the Manitoba Infrastructure and Transport (MIT)[23] from a network of gauges distributed in key places across swaths of territory. The data have to be synthesized and made meaningful by entering them into statistical models, and then they have to be translated back into information that is locally relevant. In the context of disaster management, modelling almost always lags behind the exertion of natural forces.

To be able to predict the path of an oncoming flood, forecasters build models based on flood history. In this manner, they assemble a particular idea of geographic space, an empirically grounded technical imaginary of risks and hazards. The process by which forecasters, engineers, and geographers transpose data gathered in a set of key sites into a regional model and then transpose them back again into socially relevant sets of warnings, though purposeful, is far from perfect. For example, on any given night as a flood approaches, officials might want to use provincial data to communicate which stretch of a dike could fail, which families need volunteers or the military to help with sandbagging, and which neighbourhoods should be evacuated. Sometimes in emergencies the most sophisticated infrastructures and models are less useful than, or must be supplemented by, a local farmer braving the storm to check a simple measurement stick out in the river near the barn.[24] The need to supplement and reach beyond available information and infrastructures in unpredictably dire moments is an inescapable aspect of all socio-techno-natural predicaments, i.e., predicaments that affect socially organized communities relying on technologies to protect them from natural forces (Winner 1977).

When an unfolding flood does not conform to historical patterns, as may happen more frequently because of climate change, modelling becomes even more challenging (Government of Canada 2019). Simple mistakes about how the river moves are surely less likely to happen to an engineer like James Eads, who learned the Mississippi River by feeling the currents sucking him this way and that as he walked the dark bottom (Barry 1997, 26). This kind of sense-knowledge, gained from an engineer's version of extreme fieldwork, provides a foundation that no kind of distance learning alone can ever match. Clearly, the combination of field knowledge and state-of-the-art technology will be the most effective bases for local flood control. That said, increasing technical sophistication is just as likely to exacerbate uncertainty as not (see chapters 3 and 7). The tension between field knowledge and computer-modelling–based knowledge manifests as an intergenerational tension among forecasting and engineering experts.[25]

Making an error when transposing knowledge from a map on to the territory it represents, or when integrating what you know intellectually into what you or others sense, is an inevitable part of a general process that affects flood forecasting. It is a technical glitch that can happen to anybody using any kind of map to interpret or act in any particular place. In any case, ethnographers routinely set ourselves up for tactical blunders like this, whether or not maps and their technical glitches are involved. It is the basic logistics of how we produce knowledge, our stock-in-trade. Whether we use maps, texts, or other visual technologies in our work, being in the field as a participant observer usually works better for discerning the right research questions than does distant data-gathering. This is exactly because the contradictions between embodied experience and what we think we know alert us to taken-for-granted assumptions. Exposed assumptions may induce perceptual and conceptual shifts that lead in turn to lines of inquiry that are more attuned to particular places.

My quick exchange with the woman on the promenade revealed a contradiction, a cognitive dissonance, between the more refined resident's knowledge of her habitat and mine, the newcomer's. Though insignificant, my slip-up on the meander shows how maps can tilt perception. Map-making – the art and craft of translating geophysical into representational space – may predispose those who use them to error (Lefebvre 1991). The errors are artifacts of interpretation in the making and the using of maps. And too, the stranger's error, though in this case without consequence, alerts us to the notion that maps are never neutral in technical or political senses. Like drawings and photographs, maps can reconfigure the politics of space.

Maps can reset terms of struggle and underwrite colonial relations of power (for Manitoba see, e.g., Gill [2000], Razack [2002], Grek Martin [2014], Hogue [2015]). For example, Sheila Dawn Gill (2000, 132, 142) analyzes maps as authoritative tellings, spatial narratives that perform provincial authority in Manitoba where "racism has been rendered unspeakable by parliamentary law."[26] Mobilized by powerful ideas such as the Doctrine of Discovery, maps rationalize and legalize

Indigenous peoples' dispossession and ultimately become part of cultural repertoires that make settler colonialism seem normal and right to those who benefit most readily. Indeed, Sheryl Lightfoot (2016, cited in 2020, 280) argues that the Doctrine of Discovery is an interpretation of law that continues to fortify liberal and colonial institutions and competes with "the Indigenous interpretation of 'reconciliation,' which refers to a set of transformation politics." Thus, the Doctrine of Discovery subtly undermines current Canadian efforts to define and implement reconciliation in accordance with both Canada's Truth and Reconciliation Commission (TRC) and with the United Nations Declaration on the Rights of Indigenous Peoples (UNDRIP).[27] Taking Lightfoot's insight regarding national and global-scale struggles back around into the ethnographer's neighbourhood Red River scene, the maps that triggered my shift in awareness reveal themselves as only a tiny subset in a multi-media spree of empirical yet ad hoc colonial materializations.

Maps work by generality and abstraction, projecting themselves, their "discoveries," as natural facts while in actuality, they fix, simplify, and ultimately override the heterogeneous, provisional, and contradictory nature of embodied experience. Settler maps come to be taken as givens, reference points for future action, much as I took the idea of the international boundary on the "forty-ninth parallel" dividing the "United States" from "British" (later, "Canada") as a Red River reference point. Such boundary markers tend to disappear the violence of their creation, slipping behind the more immediate requirements and pleasures of the everyday. But if one searches for them, there are powerful counter-narratives that refuse such lapses of memory. For example, Indigenous historian Michel Hogue (2015, 4) shows how Métis communities "shaped and were shaped by the border." His history opens with archival photographs of the Métis engineers who transformed what was once pure political fiction into the foundational reality of an international survey line – i.e., a linear projection into the apparently empty spaces conjoining the array of earthen mounds they built.[28]

It is not only peoples and cultures that disappear into the new kinds of everyday lives mapped into being. The glide of a pen (or cursor) smooths over the detailed life of worlds tucked away in tiny variations along ice/water's riverine edges. In maritime northern geographies, maps turn dynamic zones of sea ice formation into thin, movable lines, distorting diverse ecological and geographic realities for pragmatic law and policy ends (Steinberg and Kristoffersen 2017). In a critical expansion of riverine understanding, Da Cunha (2019, ix) finds the whole idea that rivers like the Ganga are more permanent features of the planetary surface than other moments in the hydrological cycle comes from lines drawn on colonial maps.[29]

This ethnography of provincial flood control works with multiple settler and science-based representations of rivers as ice/water bodies with shape-shifting relations to the city and to other ice/water bodies. And, yes, while one might easily ignore or forget the east-west orientation of one meander in a Winnipeg neighbourhood, there is a lesson in this detail. Setting aside the specific workings of colonialism for now, the trick to learning from maps while not falling prey to their tactical misrepresentations is to keep an eye out for the missing, the vague, and other distortions that arise when the "mind's eye" meets embodied experience, i.e., training one's attention to the intersection where the technical translates the sensible into power and authority (Jacob 2006, 11).

B. River Flow as Cultural Orientation

Whether new or old, if we introduce ourselves to riverine territory through maps or use them to argue the rightness or wrongness of perspectives or projects, we may influence or erase existing conflicts and possible alternatives for action. By their very nature, maps are a material form of hydropolitics, highlighting some bodies of water, their qualities, pathways, social meanings, and scales of visualization at the expense of others.[30] As a cultural anthropologist trained in ethnography, I keep my eye out for the particular sort of interpretive error that transpires when one travels from home to someplace far away. While this involves visual technologies like maps to navigate field sites, it does not usually involve the matter-of-fact translation of riverine landscapes from abstract, linear representations to the kind of embodied, perceptual experience of elemental relations that uncovers interpretive errors. But patterned cross-culturally, interpretive errors may have geopolitical implications, the character of which depends on the power differentials between those making and reproducing the errors (e.g., colonizers) and those who continue to experience and resist the consequences (e.g., colonized).

At this point in my narrative, however, the error at issue is a basic geomorphological one. It revealed itself to me as an unconscious cultural predisposition that came to my attention because it contradicted what I was trying to learn. A cross-cultural difference in apprehending rivers arose in the contexts of interviews with Steve Topping and Alf Warkentin, water engineer and forecaster, respectively.[31] I synthesize the subject matter in question:

> When the Red River melts, it does not melt all at once. This simple fact can create infrastructural havoc in late winter and early spring. Running *south to north, warmer to colder*, meltwater winds across the US-Canadian border through Winnipeg and eventually out to Hudson Bay. Ice jams are prone to form in the tight meanders north of the city, especially in this one spot where the course is split and narrowed by an island and bridge footings. Ice jams can block the water's northward push and force it up over the sides of its channel, causing widespread flooding across the surrounding prairie.

In the abstract, I "knew" of course that the waters of the Red River flowed northward, eventually emptying into Hudson Bay. (I'd never get that fact wrong on a test if asked directly.) But not until I listened to my interlocutors unpacking this set of ice dynamics for me did I become aware that in the back of my mind I was actively working against a predisposition to think in terms of north-to-south flows rather than south-to-north flows.[32] And not until I became attuned to this deeper zone of place-based geological knowledge did I understand how a fact can float on the surface, a symptom of what a stranger may not have thought through.

What happens when water as liquid and water as ice move through the drainage basins of major rivers and their tributaries depends greatly on this difference in spatial orientation. This is a shared form of local and regional knowledge. For this ethnographer-stranger, it was a realization that triggered an appreciation of river flow as a kind of cultural orientation. Being and becoming a body on a part of the North American continent tilting either toward the pole or toward the equator can be something felt but, without an attention-calling-sign, not known.[33] Being aware of having to overcome the predisposition to think one way about the Red and not another thus revealed what I believe to be a subtle manifestation of a basic cross-cultural difference in spatial orientation. I become aware of it only when crossing into a domain of expert-inhabitant knowledge. Does this aquatic form of cross-cultural difference come more fundamentally from intergenerational habits and implicit understandings tied to particular territories – even among non-Indigenous peoples who may not recognize this buried aspect of their relationship with rivers? Human-river orientation is part of an earthly habitus, if you will. Implicit and unknown, it operates in the sphere of unintended agencies.[34]

Something as basic as the predisposition to interpret the directionality of water flow through a landscape is mediated by cross-cultural dispositions rather than natural (spontaneous) or universal (ubiquitous); this surprised me. I had to retrain myself to think in terms of south-to-north flows whenever visualizing Manitoban rivers. Although much of what I learned through fieldwork was of a technical nature, discovering such unexpected differences in spatial orientation led me to think further about the significance of the cultural dimension of water infrastructure. A robust sense of culture that incorporates riverine earthworks could surely include a consideration of basic predispositions that orient humanimals in aquatic space. Such geo-cultural differences may turn out to be diverse and consequential, folding into and complicating other predispositions that lead us to value landscapes and peoples arrayed along a north/south valence differently.[35] In any case, I do believe that the orientation toward geomorphological space animates what Gandy (2002, 13) identifies as the "different cultures of nature" active in the world.[36]

C. Facts, Feelings, and Throwntogetherness

In her book, *for space*, Doreen Massey (2005, 131, 133) reflects on her own attempts to cross the gap between knowing and feeling facts. Her point of departure is a personal history unconstrained by the form of settler colonialism that shapes Manitoban geo-culture. She and her sister visit the Lake District in northern England, another landscape shaped and scraped by the glaciers of the ice age. They escape the city, seeking some grounding in a more natural setting. There her attention gravitates to Skiddaw, "a massive block of a mountain, over 3000 feet high, grey and stony; not pretty, but impressive; immovable, timeless." She realizes her perceptions are out of synch with her knowledge of "geological history which tells us ... that this 'natural' place to which we appeal for timelessness has of course been (and still is) constantly changing." The rocks composing Skiddaw are comparatively new, laid down in an ancient sea a mere 500 million years ago, living trilobite time, and since then subject to volcanic activity. Moreover, when the Skiddaw's rocks were laid down, the sea that laid them was in the southern hemisphere, nearer to the South Pole than to England.

Massey reconsiders her embodied relationship to what is commonly thought of as the place called "northern England" in light of glacial action, mountain mutability, and continental mobility. She extends the implications of her Skiddaw reflections with other examples (I paraphrase pp. 138–41): we know that tides move oceans, but we also know that tides move solid earth. The continental interior of North America goes up and down by about 20 cm daily. One may know this fact, but because one cannot feel it, the fact does not usually enter one's ordinary thinking. And too, the angle of earth's axis has changed significantly and somewhat chaotically over geological time.[37] Know it but not feel it? Massey argues that because everything moves, one can never return to "here." Outside of the present moment, "here" will have already moved elsewhere (due to processes such as erosion, sedimentation, and the "throwntogetherness" of place).[38] Whether consciously or not, mostly not, we craft our sense of being here rather than elsewhere through negotiation with more-than-humans that enact their state-of-being in spacetime frames dramatically differently from ours.

Reading Massey, I then realize that my discovery held the presumption of a stable "here" and "there" in place. I discover that the mode of spatial orientation I travel with misaligns my sense perceptions with the facts of fieldwork experience, but I never question the fact that when I leave the Red River Valley of the North behind and go home to the Mississippi watershed, I could safely resume my "native" (third-generation US settler) mode of orientation without incurring error. And indeed, this is a fair assumption in the historical spacetime scales in which humans tend to think and live. But this assumption exists in tension with a growing imperative that we understand ourselves and our effects as not merely historical actors but as collective geological actors.[39]

In all our glorious diversity, people tend to act in their everyday lives as if the ground upon (or climate in) which we stand is stable even when science, and the experience of disasters, tell us that the ground (and climate) are shifting. Yes, as practical social beings we have to negotiate everyday life as if it takes place, or can take place at least, in a dependable "here" from which we can go and return and a "now" whose past is behind and future in front. But as collective geological actors, some among us should have a responsibility to stretch beyond such conceptual comfort zones, in part by considering how comfort zones occupy what Leanne Simpson (2017a, 39) calls "unkept-promise land."[40]

Back in the everyday of fieldwork, in a city where floods loom over the sense of "here" and "now," I appreciate the inventive nineteenth- to twenty-first-century technical systems and processes designed to stabilize a dynamic piece of planet. I also consider the human and nonhuman sacrifices entailed in making a city stable. Like Donna Haraway (2015, 160), I write in an experimental mode captured by Jim Clifford (2013), whom she cites when she writes, "we need stories (and theories) that are just big enough to gather up the complexities and keep the edges open and greedy for surprising new and old connections." In this spirit, I return to walking the river edges near my central Winnipeg neighbourhood.

III. THE WALL THROUGH TIME IN A "POTENTIAL ANTHROPOCENE":[41] MAKING CUTS, EMBODYING LOGIC

To "decolonize" means to understand as fully as possible the forms colonialism takes in our own times.

> Dian Million (2009, 5) in "Felt Theory: An Indigenous Feminist Approach to Affect and History"

A. At The Forks: Consuming the Confluence in Downtown Winnipeg[42]

Even as they thrive on contiguity, settler river cities are founded on hardening the separation between river and land. River/land separation manifests in myriad, mundane, smaller-scale earthworks, some of which are designed for cultural as well as infrastructural functions. The "Wall Through Time," a walkway that connects riverbank strollers to market shops, restaurants, art, and performance stages on the upland, provides a minor but telling example (see figure 1.4). The wall's curve cuts off and retains a concrete-encased hill that leads to the upland, dividing dry land from the wet clay "gumbo" that melts ambiguously into the confluence of the Red and Assiniboine.[43]

Following Elden's (2010, 809) theory of territory as a political technology dedicated to mapping and control, the Wall Through Time has

1.4 Walking beside the Wall Through Time, I suggest, embodies a psychic sense of flood hazard and an unconscious appreciation for engineering river/land separation.

been "made possible through a calculative grasp of the material world," one that is "particularly [useful] for the abstract division of unknown places in the colonized world." The wall literally produces urban territory through processes of measuring, mapping, and ultimately cleaving river from land. Elden's widely applicable theoretical contribution has been disseminated in articles and books by him and others, including here, by me. But in settler colonial contexts such as Canada and the United States, there is a danger that as the theory travels through scholarship, it may further contribute to the erasure of the theories of Indigenous nations that continue to imagine territory otherwise even as they have to share it with settlers. At this historical moment, when the struggle for truth and reconciliation has taken hold of the scholarly world, Anglo-European theories are best put into conversation with Indigenous critical theory.

For example, in *The Transit of Empire*, Jodi A. Byrd (2011, xxx) suggests "imagining an entirely different map and understanding of territory and space" and offers a "diagnostic way of reading and interpreting colonial logics that underpin cultural, intellectual, and political discourses." (Byrd writes in the United States, but her analysis slips easily into Canada along with the Red River waters running from the Dakotas, across the border and into this scene at The Forks.) I offer this geo-cultural analysis of the Wall Through Time as a small step toward this new and hopeful arena of diplomatic interdisciplinary exchange.

> 1 small-scale decision x ∞ = remaking the planetary surface.
> The potential Anthropocene delineates a spacetime layer within the geological record and inspires a geo-cultural reorientation toward unfathomable planetary futures.

At some point, I figure, expert-inhabitants must have stood on that slope and decided where to make the cut, which piece of riparian hill habitat would be removed and which retained in order to make the elegant walkway.[44] Masons and city planners also invest the wall with aesthetic power, facing the vertical concrete with Tyndall Limestone.[45] To make this possible, miners must have dug down into the early Paleozoic (450 million bp) to find the world of trilobites that once burrowed into the mud bottom of a warm inland sea. Although the trilobites died after continents moved and climates changed, their empty burrows persist as fossilized texture in the creamy stone facing. Lifted up and out of a downriver bank of the Red River near Selkirk, the wall's thin slice of stone bears the zoo-architectural beauty of trilobite extinction.

B. *Making the Familiar Geologically Strange*

> The relocation of a horizontal Paleozoic trilobite habitat into an Anthropocene vertical.

> "But why," asks a hypothetical future paleontologist looking back at our everyday lives from long ago?
> Only humanimal engineering could explain it.[46]

Two bands of blue tiles, parallel to the ground, cut through the Tyndall Limestone facing. The high band marks past floods, maybe 1826, maybe 1950, inducing a disorienting contrast with the everyday.

Ghosts of swollen rivers rise up in strollers' sunny-day imaginations. (I extrapolate from my own experience and hypothesize from informal observation.) Where the high band hovers way over even the tallest among them, strollers may project their imaginations into a situation in which their feet would not have been able to touch solid ground or even the sucking gumbo.

This is my interpretation: Physical proximity to the wall's high blue band nonverbally communicates the extent of extreme physical danger that flood control engineers have quelled. The wall's blue-tiled referent kindles concern for water's potential to rise from converging currents. Such concern feeds the rationality of the cut, the separation of land from river, and, more generally, the implausible centring of the city in the intersecting flood plains of two strong rivers. And so, moving back and forth along the wall, strollers' bodies register the danger that existed before the modern flood control system afforded city-dwellers a sense of water security. Modern infrastructural habitus – i.e., a human-river orientation that integrates a strong engineering response – can be cultivated in the design of civic space.[47]

But there is more. Informing and at times intervening in this book's geo-cultural analysis of flood control is the knowledge that the very system designed to keep the city safe imposes ongoing colonial relationships in the homelands of Indigenous peoples who have their own legal cultures, including river management (Napoleon 2007, 139). To introduce the adamant yet somehow obscured nature of this historical fact, I decode the wall to show how its symbolic infrastructure represses local and Indigenous knowledge with an alibi.[48]

Upon its Paleozoic surface, a series of plaques and panels narrates progress toward the industrialized city, its makers still innocent of climate crises to come. The visual narrative pushes the iconic tipis and names of Indigenous river/land events back and down toward deep time. In infrastructural terms, the wall narration misrepresents pre-1950 settler infrastructural habitus, a "land culture" founded on river-control that appears to rise out of and supersede an "amphibious" infrastructural habitus adapted to river being and becoming.[49] I say misrepresents because Indigenous people have certainly not been left behind in the city's past (Tomiak et al. 2019). Such misrepresentations are constructed and influence what is taken to be common sense.

I attend closely to the first panel nearest the riverbank because it incorporates the model of earth's formation in a series of six layered geological-to-archeological events leading to present-day Manitoba.

Topped off with The Forks Redevelopment in 2000, the layers get longer and longer as time deepens. Each layer in this framework is filled with time-stamped icons and names (see end notes for details):

<u>A Settler Chronology of Known Events at the Forks</u>

The Railway (1870s–2000)[50]
Formation of the Province (1800–70)[51]
Fur Trade (1720s–1820s)[52]
Native Bison Hunting (time immemorial to 1720)[53]
Lake Agassiz (11,000 to 6000 bp)[54]
Glaciers (millions to 11,000 bp)[55]

This quasi-scientific schema, designed for the passing appreciation of strollers, naturalizes the colonization of Indigenous people and rivers. The panel is meant to celebrate and geologically situate the city's accomplishments, but in so doing, it disguises the intentionality and cruelty of urbanization. In this matter-of-fact way, geo-logic affectively shapes geo-culture in the nooks and crannies of civic space. There are ongoing scholarly challenges to this all-too-common geological imaginary.[56]

In her book, *A Billion Black Anthropocenes or None*, Yusoff (2019, 8) asks "How is geology an operation of power, as well as a temporal explanation for life on the planet?" The wall's geo-cultural revelations fit within an answer to Yusoff's broad question. First, the wall's particular way of visualizing geological time derives from studying material traces of events accreting into the earth's crust, newer on top of older, epochs on top of epochs, eons on top of eons. Enacting "geontopower" (cf. Povinelli 2016) and "settler geology" (cf. Schmidt 2020) the Wall Through Time fits cozily within racialized models of progress, civilization, modernity, development, and disciplinary knowledge (cf. critiques by Fabian 1983, Razack 2002).[57]

C. Archive and Anthropocene Dialogue

Taking an objective, long, linear, post-Anthropocene perspective, geoscientists calculate that rivers probably have a future on this planet despite everything cities have already done to them. Unlike the carbon cycle, "which will take tens to hundreds of millennia to regain equilibrium once human pressure on them stops," fluvial systems "can rapidly (from a geological perspective) revert to their original patterns" (Williams et al. 2014, 62). Once it expands into geo-culture, however, the objective, long, and linear perspective of geoscience obscures crucial cross-cultural modalities of humanimal–river relations; its simplifications erase too much survival know-how. To move into geo-cultural work, the Anthropocene has to be rebuilt from the ground up, i.e., cut down to riverine size and opened up to place-based histories. Unlike the Paleozoic trilobites raised up from deep time strata to beautify the Wall Through Time, Winnipeg's inhabitants are very much alive: our geological strata are still in play.

And so, to return to the book's beginning, if we take seriously the idea that the "impulses of rivers and human collectives (cities) combine as they carry on being and becoming, persisting in distinctive yet entwined embodied form," then life on earth as we know it could survive a potential Anthropocene along with rivers. To that end, it's worth trying to understand the twists and turns of one river city. I gather a collection of key fragments from official provincial history that show how (post)colonial law, engineering, and geoscience have reworked Winnipeg's terrain, initiating and installing a recodification of river/land spatial relations. I put this bare-bones archival sample, a pastiche of territorial precedents, in dialogue with geoscientific data characterizing the "human footprint" on river systems. Gathered by Syvitski et al. (2020), the authors include members of the Anthropocene Working Group (AWG) who argue in support of officially recognizing a potential Anthropocene epoch that would end the Holocene epoch in 1950. I call these juxtaposed event-lists the "Archive and Anthropocene Dialogue" (see Appendix I).[58]

The "Archive and Anthropocene Dialogue" clearly shows that Winnipeg's regional-scale, flood control history lends substance to the rationale for a planetary-scale 1950 cut-off (after which the modern system was built). The dialogue may thus deepen appreciation for river cities as collective human actors affected by and causally implicated in the precarious intersections of our historical and geological times. At the same time, the archive clearly shows the intentional and systematic work by settlers and their descendants to separate land from river and separate Indigenous peoples from their land/river homelands. By putting colonial settler history and geoscience into a condensed conversation structured by historical floods and infrastructural development, the Appendix shows how the Anthropocene unfolded through forms of "progress" enacted through legalized Indigenous oppression.

When rivers encounter the city, I suggest, they encounter a human collective, but while rivers may not distinguish among groups and the different responsibilities they hold for the urbanized earth's state-of-being, the archival details force attention to assaults on Indigenous lifeways sanctioned by the Doctrine of Discovery, among other ideological devices like *terra nullius*. These assaults are as much a part of Winnipeg's progress into modernity as the railway and the flood control system. Indigenous intergenerational experience of settler colonialism certainly puts the apocalyptic urgency of human-caused Anthropocene, and for that matter, of climate crises, into perspective. As Whyte (2018) clarifies in his analysis and review of Indigenous writers, activists, and political representatives, these are not *the* intergenerational crises of our times but rather more terrible iterations of conquest and colonialism.

What if Indigenous people and philosophies had had – and were to have – a stronger balancing influence on Manitoba's development? For example, in her essay on precolonial diplomacy, Simpson[59] defines *mino-bimaadiziwin* (the good life) as "a way of ensuring human beings live in balance with the natural world, their family, their clan, and their nation." This central Anishinaabe concept may well share some conceptual terrain with ecology even as it contrasts starkly with the calculated sense of balance enacted by the engineering-law alliance that governs flood control (see chapter 5). If Anishinaabe culture experts were interested in sharing their understandings of *mino-bimaadiziwin*, it might orient cross-cultural negotiations and help to open flood control governance to the aims of "ecological democracy" (Shiva 2002, 15), to practising "collaborative consent" in the realization of UNDRIP (Simms et al. 2018), and in thinking with "critical settler cartographies" (Fujikane 2021).

This ethnography builds a framework and terminology that pulls the science-based technicalities of flood control and law into an arena of pragmatic imagination in which such questions can be asked and Indigenous theories applied. It clears a space for livelier conversations in riverine governance.

IV. TRACES ON TREES:
PAST FLOODS AS LIVING HISTORY

Colonial imperatives had been set in motion long before fieldwork. The old maps, despite their faults as arbiters of reality, offer this ethnographer clues about a time when the cuts separating land/river were fresh. I can see details relevant to my neighbourhood walk on the 1874–75 survey map I found in the Manitoba Archives (see Appendix I for details). From the Lyndale Drive in 2014, the land defined by the river bend looks like a collage of all three map categories: farms, marsh, and woods. Looking across the channel, the woods would have extended as far as my eyes could see, way past the protected remnants in present-day Assiniboine Park and Assiniboine Forest. As I learn from Bill Rannie, the geography professor, if I were walking along the nineteenth-century landscape where the present-day promenade is, I would be lower, closer to the water. The channel would be narrower because it would not yet have been widened by the many bank failures to come.[60] The roads, houses, and infrastructures would be different. Yet some feelings, some aspects of sensual experience, of the powerful aesthetics that bind today's dog-walkers to yesteryear's inhabitants might resonate across time. The pull of the river, its powers of orientation and affect, persist.

One afternoon on the Lyndale Drive, the dog-walkers split off along a gently sloping gravel path to enter a greensward adorned with trees and furnished with little garbage pails and plastic bags for poop. Thoughtfully placed in pretty spots facing the river, freshly painted memorial benches invite contemplation, conversations in mostly French or English, or consultation with Internet screens. The park opens up at the end of the river's east-west stretch as it bends toward the north once more. Away from the river, back across the lawn toward the neighbourhood, the high diked pavement for cars and bikes visually divides the park – a floodplain buffer that it will not protect – from residents' homes, which, if all goes well, it will protect. Dikes had to be located well back from the river's edge because of the danger of unstable banks (Greater Winnipeg Dyking Board 1951). This linear high-elevation dike, which is also a vehicular road, pedestrian promenade, and infrastructural platform, repeats throughout the city and its surround, materializing a sense of order, continuity, and dialogue with the wet prairie-city landscape. At least this is how it seems when times are normal: in everyday scenes, when the river behaves, when the things of the world are in place, when you have a home and it is within a protected zone.

One day, I bumped into my next-door neighbour walking his dog in the park and asked about the Great Flood of 1950, the last great flood in living memory that drowned a central swath of Winnipeg.

This happened before he was born and before the promenade had been transformed into a primary dike. The flood is part of his family's oral history. His father had shown him the inundated part of the neighbourhood called the "flood bowl" and how the bark on the lower part of the trees that had stood in the floodwaters was lighter in color than the rest. By drawing an imaginary line among the trees connecting them at the height of these registration marks, the affected area emerges from the past. By virtue of its material after-effects, the flood bowl – one of hundreds if not thousands of flood bowls dotting the all-encompassing flood bowl that is Manitoba – hovers like a familiar ghost in the park's present.[61] Wanting to characterize this neighbourhood flood bowl more precisely, I return with a compass.[62] I confirm that the river broke through its banks (and, in 2011, broke off sections of the bank) precisely where it turned the bend to resume its northern course. It's as if the river holds its south-to-north flow as an ever-present impulse, even when at certain points the terrain forces the channel to veer west or east.

Meanwhile, the future of the homes along the Lyndale Drive promenade has been made much more secure by the combination of this particular piece of the primary dike and the larger system of which it is a part. Especially important is the Red River Floodway, built in response to the spring 1950 flood. Coming on line in 1968, it stops excess water from entering the city from the south, taking it around thirty miles to the north before dumping it back into the Red's channel. The key structure of post-1950 mighty infrastructural projects, the Floodway saved the city in the 1997 flood when the Red River rose up and spread into a sea (or lake). Before retreating back into its channel, it drowned out many of the towns and farms between Grand Forks, North Dakota, and Winnipeg's southern suburb Grande Pointe.[63] Because they are among the inhabitants of the protected centre, however, the dog-walkers on the Lyndale Drive occupy one of the privileged places in the provincial landscape. They are reminded of this privileged status whenever a major flood overwhelms the homesteads outside of the protected zone.[64]

Public and private collections of black and white photographs keep the intergenerational memory of the 1950 disaster alive. They transport those who look at them into the scenes of disaster. Registering the uncanny transformation of familiar streets and landmarks, neighbourhood by neighbourhood, photographs link past events to present and future, folding floodplain reality into the spine of cultural identity. In *River Rampant*, a pictorial record originally published to raise money for the Manitoba Flood Relief Fund of 1950 at $1.00/copy, one page features events at Norwood Bridge, a vital link that then, as now, connects the central business district to the residential neighbourhoods on the east side, including the neighbourhood that is now bordered by the Lyndale Drive promenade.[65] The photo shows an assortment of old-fashioned vehicles chained together in a line and loaded down with people and supplies. A tractor in the foreground anchors the scene, connecting the viewer to the men in coveralls, plaid shirts, and caps being pulled through as the water reaches for the tops of the wheels and laps at the protruding front lights. No other people are visible in the scene.

Sticking up from the surface, posts, wires, a fire hydrant, a sign with letters or numbers and an arrow, and a curved pile of sandbags compose a guide to the urban fabric that lies beneath the surface. The photographer somehow shoots from above, facing the supply chain. The text accompanying this photograph explains that as the river rose, people kept close watch on the bridge as they piled up sandbags. Fearing that the weight of debris heaved by the water against the foundation threatened the structure's safety, workmen placed ladders against the foundation's sides. They climbed up to remove debris and load it onto waiting trucks, eventually removing fifty-five truckloads. The sandbag dikes stayed solid, but the water seeped through on the approach from the city, which triggered the improvised ferry system shown in the photo. In the end, the bridge had to be abandoned. The city evacuated about 100,000 people. The evacuees' escape entered into the oral histories of families and towns far and wide who took them in, introducing unexpected and sometimes unsettling encounters between hosts and evacuees with different class and ethnic identities.[66]

The photos, their captions, and the stories of the elders who survived 1950 remind inhabitants of the chaos that rivers bring and affirm the value of infrastructural investment. They capture moments when earthworks shift from the background of everyday life to the forefront of elemental battles, affectively transporting those who see and listen to a prairieland texture transformed by rivers in pursuit of glacial destinies.[67] More than the Wall Through Time, which has similar geo-cultural functions, visual and oral documents of past extreme events remind people how flood-fighting relies on social bonds, especially among heroic strangers who enhance the city's resilience by sandbagging home after home while seeking nothing but hot coffee and sandwiches in return. They also dramatize the rationale for perpetual investment in flood control, legitimizing and expanding the power of the institutions responsible for infrastructural operations and emergency management and the economic vitality of the private engineering firms that carry out the construction under contract. Thus, pointing backward and forward in time, the disrupted landscape and its representations enhance belief in collective action by communities, governmental agencies, businesses, and the military.[68]

The Red River flood of 1997 was far more ferocious than the flood of 1950. Had the provincial system not been in place by then, it is fair to say that Winnipeg would have surrendered to the Red and would not have become the thriving city that it is today. The 1997 flood justified the investment in the modern infrastructural system as well as subsequent improvements, including the expansion of the Floodway. To everyone's relief and amazement, the Floodway held off the destruction they saw unfolding as the Red Lake rushed slowly but inexorably across the valley until it was finally turned away by the Floodway's control dam. But the Floodway reached the capacity it was designed for; one more hard rain as the river crested and it would have failed.[69]

If the river had gone any higher, the engineers would have had to blow out bridges, which would otherwise have acted as dams that would widen the flood. The government was making plans for the evacuation of 700,000 people, just in case.[70] Once the emergency passed, design flaws could be fixed, and city inhabitants could confirm their faith in their collective power to hold off the worst the Red had to offer and affirm the city's investment in and reliance upon earthworks. And so, 1950 and 1997 floods – the before-and-after markers of engineering prowess – became a popular frame of reference for thinking about the city's power to control the Red.[71] But then, along came the peculiar floods of 2011 and 2014, when attention shifted to extreme storms in the west and the threats flowing in along the lower Assiniboine. Whether Winnipeggers fully embrace the concept of climate change as cause, the shape of uncertainty associated with floods is changing. Winnipeg's amazing feat of flood control engineering may have been designed with geophysical assumptions that may not outlast the twenty-first century. Now what?

V. THE WINNIPEG MODEL OF FLOOD CONTROL: READY FOR EXPORT? AN ETHNOGRAPHIC DOUBLE TAKE

Like most cities, Winnipeg has certainly not conquered river-nature or, it could be argued, has not even made serious headway in recognizing or adapting to its version of climate change. But it has pulled record-breaking storms, floods, and icy blasts into the scope of human negotiation and accomplishment. I admire the city's geo-cultural achievements, how it grounds itself in extreme territory, how its forecasters and engineers prepare the city for what may come, while respecting prediction's limits in the midst of emergency. It is a good place to study flood control as an important form of geo-culture, which, as stated above, I define as the combined logistical domains of professional and everyday life that persistently work to imagine and translate the technical into the territorial and fearful emotions into rational systems of control.[72]

One thing is sure: there is no such thing as perfect infrastructure. How cities deal with inevitable harms resulting from trying to do the best for the most people is the social justice project that must accompany sustainable infrastructural development in democratic societies (Kane 2012). To condense and clarify contradictions that arise from infrastructure's imperfection, I cast the technical aspects of the Winnipeg flood control model in both appreciative and critical lights. On the appreciative hand, in "Political Engineering," I draw on expert understanding to show how the flood control system works to separate inhabitant living and working spaces from excess river water in geo-political spacetime.[73] On the critical hand, in "Spatial Justice," I reveal the colonial legacies embedded and invisibilized in calculations, structures, and routines that justify systemic separation of protected, unprotected, and unnaturally harmed. Some key infrastructural inequalities can be traced to intentional actions of prior generations through the 1970s. Many of these actions would probably now be illegal, yet they have been allowed to melt into the sphere of unintended agencies while retaining material form and impact.

A. Political Engineering

Winnipeg is a successful intergenerational experiment in how to live comfortably with environmental extremes. A democratic provincial government is committed to keeping the city dry, despite the hazardous nature of sustaining the political and economic engine of the province in the middle of a flood bowl. The city may be in the forefront of flood control engineering worldwide and could provide a model that other river cities could adapt in part or whole. For example, singular features, such as the amazing Floodway that moves floodwaters from the Red River around the entire city, could be adopted by cities with plentiful flat, slightly tilting space for shunting water away. Adopted as a whole, the Winnipeg model would work best in northern latitudes with lots of ice; low population density in the immediate surrounds; a hydrology uncomplicated by tidal surges, rising sea levels, or high mountains; and floods that come from far away, giving plenty of warning.[74]

In the continental interior far from any ocean's moderating force, the city's infamous winters hold tenaciously onto atmospheric cold descending from Arctic latitudes. As a result, managing ice is a central design feature of Winnipeg's flood control system. The Forks, at the confluence of the Red River (legacy of Lake Agassiz) and the Assiniboine River (its wily predecessor), sits in city centre, once a small settlement that expanded outward from entwined and fortified riverbanks. Every spring, while ice struggles against liquid water, the two rivers threaten to enact minor versions of the Outburst Quartet. But with flood control, the city is protected.

The infrastructural transition from overwhelmed city to safe city is core to Winnipegger identity. Extreme floods of the past mark turning points in urban history and fill annals of folklore, so it is not surprising that inhabitants tend to appreciate experts who keep the city safe. Though not immune to conflict over operating decisions, especially in emergencies, Winnipeggers I speak to generally seem to value the knowhow involved in designing, building, operating, and maintaining the flood control system that keeps their homes and businesses on terra firma. This general consensus can be considered a geo-cultural feature that contributes to the success of the Winnipeg model.

In emergencies, engineers may find it necessary to make operating decisions likely to threaten one or more *infrastructural outsiders*, whom I define as groups whose homes, businesses, and farms are subject to extra flood damage by virtue of their proximity to the wrong side of control structures. The key difference separating them from the protected majority is defined by where their home or business is located in relation to flood control structures and operations. Is there an implicit space-based cultural and political identity at work here?

Like ethnic, racial, religious, or sexual minority status, the identity of infrastructural outsiders is animated by, yet exceeds, particular events.[75] (E.g., the identity of a family who lives south of the Floodway control structure may become most salient in a particular flood emergency that threatens them with backwash, but the space-based sense of precarity stretches from past to future floods. Where they are, I suggest, becomes part of who they are.) Unlike other minorities, however, the identity of infrastructural outsiders remains unnamed, submerged in the intersections of identity and community. As far as I can tell, infrastructural outsiders do not (yet) self-identify and organize politically across racial and class groups. As such, the meaning and usefulness of infrastructural outsiders, as a geo-cultural identity, emerges out of ethnographic necessity. After all, so far, the creation of infrastructural outsiders seems to be inevitable with any major infrastructural system.[76]

And so, to be as fair as possible, the language of legal protocols regarding mitigation, compensation, and contestation relies on flood impact calculations designed to be flexibly applied to any affected community. Authority to make decisions that likely entail negative repercussions for infrastructural outsiders gravitates up through provincial hierarchy, from MIT's back offices to the provincial premier's office, and, when emergency threatens to become disaster and the military has to be called in for logistical support, to the nation's prime minister.

Building on the concept of political ecology, which recognizes that politics shapes ecological habitats along with nature, I call this kind of decision-making arena *political engineering*, which recognizes how politics shapes the engineering of urbanized ecological habitats.[77] In post-flood event phases of political engineering, policy-makers endeavour to separate natural from operational causes of assessed flood damage. They define nature as a hypothetical level of river or lake at flood stage. Using the actual measure of flood stage levels, engineers ask: what level would the river have reached at a home, farm, or business site if no flood control system had been operating? Natural cause, so calculated, is distinguished from artificial causes, i.e., damages caused by operational decisions. Such technical calculations, written by engineers and mandated by law, determine which households receive compensation for flooding (those damaged by operations) and which do not (those damaged by nature or classified as ineligible for other reasons). In such ways, decision-making procedures are always accompanied by technical traditions designed to measure social justice as lost hydrological balance.

That the flood control system inevitably prioritizes safety of the urban majority and, the corollary, that the system unintentionally but predictably intensifies hazards for minority-outsiders, is, alas, an unfortunate fact of life. As the chief engineer at MIT explains, "The water has to go somewhere."[78] Thus, the managers of the Manitoban flood control system aspire to fairness within rationalized limits.

B. Spatial Justice

Should other, less prepared river cities facing flood extremes for the first time in their histories seek models of successful flood control systems like Winnipeg's?[79] I argue that at this critical juncture, when cities seek pathways toward climate change adaptation and resilience, it is crucial that before export, inhabitant-experts must interrogate the unintended harms accompanying the technically advanced flood control systems designed to protect them from the elements.[80] Attempts at transferring engineering practices and allied law and policy architectures should not be treated as straightforward technical matters that take account only of such things as physical terrain, expertise, financing (important as these are and even if so-called stakeholders are doing the accounting). Even with the best intentions, dependence on model technical systems without benefit of pluricultural critique reinforces and disseminates crimes and errors sedimented in the material foundations of cities. As with climate change action more generally, flood control initiatives must move social justice for infrastructural outsiders to the top of priority lists. And too, because humans share the web of life with more-than-human beings and things and flood control infrastructure always has simultaneous impacts, social and environmental justice are best considered together when designing core interventions.

There is, of course, always resistance to change, especially if the status quo satisfies the majority needs for safety or if it requires inhabitants to relinquish accustomed creature comforts. I get it, I feel it myself. But the fact of the matter is, discrimination and injustice can arise out of good intentions, such as building and operating a flood control system that is a technical black box to most of those who benefit. And so, urban inhabitants and planners must actively look at how default legacies of colonialism are built into the way flood control produces space (cf. Lefebvre 1991).[81] And beyond material structures and operations, it's important to understand how streams of knowledge and expertise flow into this alliance of engineering and law grounded in geoscience. How to think about this?

Here, I turn to Philippopoulos-Mihalopoulos's (2015, 66–7) approach to what he calls *spatial justice* in which space can only be produced through law and law through space. Thus, the provincial flood control system built around the confluence of the Red and Assiniboine Rivers can be interpreted as an expression "of the material nature of law itself," a *lawscape* as much as a landscape (see figure 1.5). A lawscape has actual forms: for example, offices of experts and officials, texts specifying operational and decision-making protocols that actively link to living rivers like the Assiniboine by means of material structures like the Portage Diversion (e.g., Scene #2 at the top). A lawscape, he writes, also has its "potential, latent form that might emerge at any point." In this ethnography, I analyze flood control as lawscape, as actual and potential forms. Linking intentional earthworks to the sphere of unintended agencies allows me to consider the logistical powers that the state borrows from water in the same frame as the submerged material history of colonialism and, too, the technical traditions of forecasting in relation to increasing unpredictability of climate change.

Manitoba's lawscape sacrifices infrastructural outsiders to protect the city and, insofar as is practical, compensates them for their sacrifices. It is worth contemplating how this situation has come about. Spatial inequalities of emergency floodwater distribution – an actual form of control – arise quite directly from designs drawn on nineteenth-century maps that redefine the river/land interfaces in Winnipeg as fixed or fixable. After the Great Flood of 1950, the basic nineteenth-century settlement design develops a privileged engineered core. In tandem, settler-culture redefines river/land interfaces in northern hinterlands as nature subject to manipulation as needed.[82] By the time civil and Indigenous rights activists began to clearly move the moral compass that governed colonial alliances between engineering and law, the foundations of modern flood control had literally been set in concrete. So although activists have been able to rupture many unequal protocols and practices in the lawscape, flood control (and, to an even greater spatial extent, hydroelectricity production) persist as mundane forms of "slow violence" punctuated by disasters (Nixon 2011).[83]

Those who continue to struggle against slow violence may be repeatedly invisibilized by the very discourses of amelioration, such as the proliferation of news images of victims in the *Winnipeg Free Press*, presumably intended to help. This is particularly the case with infrastructural outsiders in Indigenous territories. Protocols for harm compensation and mitigation assume a uniformity of law. This assumption contradicts the diversity of Indigenous legal cultures. The contradictions trouble the status of negotiations, delaying critical support for Indigenous infrastructural outsiders.[84] Of course, this situation is neither new nor unique to the domain of flood control, river engineering, or water resources more broadly.[85] Even as Indigenous rights are increasingly recognized by the majority of Canadians, flood control imperatives, structures, and operational protocols stubbornly persist.

In light of the positive and the negative, this ethnography presents a geo-cultural framework and terminology designed to explore river-city flood control as an arena and method for reimagining collective survival. I unpack flood control's empirical bases and the rules by which city and province realize their commitments to fairness. Cultural analysis of the categories used to operate and evaluate a system with proven technical success can, I suggest, realign the nature of elements and how they are engineered to put social justice at the core of flood control. Why not? Winnipeg's inventiveness offers a kind of language, a way of interpreting hazard and uncertainty that may reassure, or at least help to clarify, the perils and possibilities of re-engineering a city's relationship with its rivers.

VI. THE CHAPTERS

Each chapter is an experiment designed to yield unexpected geo-cultural insight through a set of empirically grounded flood control stories. Along the way, the writing moves against the grain of officialdom. Techno-science is freed to roam around a more expansive sense of geo-culture with the assistance of a set of widely applicable new terms: sphere of unintended agencies,[86] blended habitus, infrastructural outsiders, internal legal frontier and political engineering, spatial justice.[87] I also discover the cultural significance of avulsion, a hydrogeological term for riverine channel-jumping. Together, the terms mobilize dynamic tensions inherent in un/intentional action,

1.5 The actual confluence in the muddy heart of the lawscape, a.k.a., The Forks.

in/visible causes and effects, material/symbolic forms and functions, cultural/political aspects of science-based earth/works, and the intimacies of space/law.

This is an unconventional ethnography; there are not actually a lot of people in it. The principal actors are rivers, lakes, and a flood-controlled city.[88] Engineers, other experts, neighbourhood folk whom I meet during fieldwork, and geoscientists who arrive in the book through their published writings appear as interlocutors to help me tell the story of the city's riverine edges and surrounds. Ideas of anthropologists, geographers, science-technology-society (STS) writers, feminist philosophers, Indigenous studies scholars, and historians of science appear here too, contributing to an open and dynamic transdisciplinary space.

The chapters loosely follow the chronology of Winnipeg's extreme flood events. This overview highlights a key theme from each.

Chapter 2, "Ice/Water on the Move," celebrates roaming and serendipity in the making of empirical geo-cultural texts. In it, I tell a flood bowl origin story in which geoscience is both source and subject. I write with an erratic spirit in mind because, as the first nineteenth-century geologists discovered, bodies of great boulders (or great ideas) torn by glaciers from one place (or discipline) and left in distant elsewheres can lead to world-changing knowledge. And too, erratic clues to deep-time events can be added into twenty-first-century computer models and contribute to the making of plausible yet hypothetical events on the grander, almost unimaginable scale (for a freshwater body) of the Outburst Quartet. The stories of erratic collisions and outbursts I pull from the annals of geoscience stretch and merge deep-time-future-horizons to set the stage for chapters to come and to contribute to the broader discussion of the "immersion of humanity into geological time (Yusoff 2013)."

Shifting downscale into the contemporary flood bowl, chapter 3, "Systems and Seasons," begins confidently in a mid-to-late twentieth-century system designed with expectations that the largest threats will come north across the US border with the Red River in spring. The Great Flood of 1950, which triggers the construction of the modern system, is the chapter's departure point. The chapter ends with stormy harbingers of climate change: queer summer rain-fed floods of the twenty-first-century Assiniboine River coming from higher terrain across the western prairies.[89] This chapter shows how forecasters and engineers continue to entwine routine operations in synch with seasons. Action quickens as swollen rivers approach and threaten to drown the city. They condense data on swirling inter-basin dynamics in the form of official, mass-mediated communiques. These gather coherence around measurements of river level at a crucial point called Number-Feet James, where the Red and Assiniboine Rivers come together and flow across the intersection with James Avenue in the Exchange District. Inhabitants rely on this piece of flood control folklore. They can interpret its significance because they have mastered intergenerational (settler) historical knowledge. They relate the information coded in this abstract index to situated details in their own habitats, such as how high the river got up the front steps of the house when Number-Feet James was so high in a previously experienced flood. For as it turns out, in urgent spacetimes of impending disaster, routine operation of digitized monumental systems may give way. When push comes to shove, finding safety relies on security dispositions cultivated and enacted at home, in neighbourhoods, in art forms, and for old-time native forecasters, from walking observantly in creeks during storms. And so, too, humble sandbags enter this chapter's ethnographic scenes as the quintessential images of nomadic cooperation and life's throwntogetherness (cf. Massey 2005).

Chapter 4, "Falling out of Synch," highlights scalar mismatches that illuminate vulnerabilities of the flood control system as a provincial territorializing project. The chapter highlights the system's reliance on blended habitus, which I define as floodplain dispositions shared by inhabitant-experts and informed by a blend of technical expertise and folk experience.[90] A comparison of two neighbourhoods in the centre of the protected zone but on opposite sides of a river bend shows the differential impacts of public and private property regimes on local water governance. The chapter also highlights a critical moment in the 1997 Flood of the Century, when an inhabitant-expert's timely realization that the Red River was about to destroy the Floodway's design assumptions and make an end run around its protective wings into Winnipeg. His insight triggers a herculean effort to come to terms with climate change on the fly.

Chapter 5, "Furies of Wind and Wave," begins with the sense of balance that lakes inspire on calm days and ends with calculating just how much injustice infrastructural outsiders can bear or be compensated for. The focus is on Lake Manitoba during and after the flood of 2011. Winds whip up gigantic waves that destroy a swath of the lake's southern end, but winds, according to a law of averages, drop out of compensation equations. On the far north end, where Lake Manitoba flows into Lake St Martin, First Nation communities receive the brunt of all excess. Wiped out, their members live in Winnipeg hotel rooms for years afterwards. Hydrological balance, for whom? According to whose definition of balance? Ethnographic scenes in this chapter are based on conversations with persons from infrastructural outsider communities; a site visit to an affected community in the southern, Lake Manitoba lakeshore; and an analysis of a documentary film and newspaper articles about the Lake St Martin Anishinaabe (Ojibway) community on the far northern end. Analysis reveals how engineers and lawyers build an internal legal frontier, which I identify and define ethnographically as a functional but legally obscured boundary that is within the province but outside the zone of provincial protection. The internal frontier, I suggest, demarcates a geo-cultural cut that divides the fully or somewhat protected from the dispossessed.[91] This chapter reveals the way habitats and communities on the far side of the cut are punished by a system-wide reliance on a colonial-era canal which, like the Portage Diversion, expands the Assiniboine basin by artificially linking river and lake bodies.[92] Unlike the system's concrete structures, the internal legal frontier will leave no traces for paleontologists to find in potential Anthropocene futures. However, as a critical mode of visualizing the spatial logic of settler colonialism twenty-first-century-style, internal legal frontiers have wide but disguised relevance for political engineering in and beyond Manitoba. Like infrastructural outsiders, the concept of internal legal frontier is an ethnographic finding that sheds light on (un)official geo-cultural shenanigans.

Chapter 6, "Sunlight Machines in Floating Ecologies," explores another dimension of human–water relationships that exists in, beside, and outside the ambit of the flood control system, enabling it and affected by it, if only indirectly, unintentionally. Widening the circle of consideration to include other kinds of kin and caring, this chapter ventures into microscopic life forms and lacustrine architectures (cf. Haraway 2015; 2016). Inspired by Barad's (2012, 53) problem of how "matter makes itself felt," I trace the contours of a collaborative space through which scientists collect and convert living and dead diatom bodies into visual images. I show how diatom bodies – buoyant sunshine machines and dead particles in geological cores dug from the flowing waters and dark underbellies of lakes – make themselves felt as botanical objects and geological evidence. And I sketch a path through which the taxonomy and ecology of diatoms gets taken up by Manitoban geoscientists who empirically reconstruct the post-glacial relationship between Lake Manitoba and the Assiniboine River. This piece of knowledge helps to rationalize the Portage Diversion as a human-made version of a once-natural link. The roundabout journey through the study of lakes (limnology) and lake-beings in chapter 6 touches down in a transdisciplinary, transhistorical, multi-scalar space that brings geoscience into play with natural history, botany, biology, and paleoecology.

In chapter 7, "River-Trickster on TV," the flooding Assiniboine of 2014 triggers a state of emergency. The last thing the province wants is an "uncontrolled breach" that could evade the control system and lead floodwaters into Winnipeg. So the province prepares to enact a "controlled breach" in a bend at Hoop and Holler, a settler farming community. They breached the Hoop and Holler dike in 2011, and they're getting the bulldozers ready to do it again. It is not an easy thing to announce to those who have so recently suffered. The premier, who would make the ultimate decision to sacrifice this infrastructural outsider community again if and when the time comes (in 2014, it wouldn't), convenes a televised press conference. He is flanked on one side by two MIT engineers who are in charge of forecasting and operations teams and on the other by the ex-military, civilian head of emergency measures, who directs on-the-ground military support. These Manitoban authorities, four men calm and serious in demeanor, attempt to inspire confidence while negotiating uncertain danger. I analyze the press conference transcript as a chronotope (Bakhtin 1981), a carefully worded representation of efforts to alter imbalances in river-city states-of-being and thereby to avert dangerous change and affirm stable continuities (not unlike the Wall Through Time).[93] I devise an experimental method of visual-textual analysis that exercises Marilyn Strathern's (1992, 3, cited in Helmreich 2003) insight that rather than measuring continuity and change, we might show how each depends on the other to demonstrate its effect. The experimental results provide evidence that, in fraught moments of significant geo-cultural crisis, the local TV press conference is a crucial genre for performing water governance-as-political-engineering.

Chapter 8, "Avulsion!," presents an unanswerable problem revealed by a new visual technology. In every flood, there is a possibility that the Assiniboine River will rise up out of its channel and sweep across the flat prairie. No news there. Theoretically, geoscientists and engineers know that the Assiniboine might not return to the same channel it has flowed through during human history. However, LiDAR, an aerial survey method using laser instead of radar, has recently produced visualizations of six or more paleochannels that the Assiniboine has flowed through since it encountered Lake Agassiz at the end of the Pleistocene epoch (MIT 2013b, 7). The new geovisualization gives clear shape to a fundamental but formerly implicit uncertainty in the relationship between river and city. The Assiniboine could avulse with any big flood, and if and when it does, it might not sustain its historical relationship with Winnipeg or its flood control system. Climate change may be activating a seemingly mythical hydra-headed river monster, but visual technology makes it seem quite real. In the last chapter, I mobilize this sense of the techno-uncanny I find along the ethnographic byways of flood control.

And now, the text moves into ice age end through the eyes of nineteenth-century geoscience ...

2

Ice/Water on the Move (Prehistory)

I. DATA COLLECTION AND THE CORPUS

A. *Learning from Erratic Encounters*

Nineteenth-century geologists collect data as they walk from point to point measuring ice/water's actions on earth's surface. They interpret and craft deep time data for future global transmission. Their visual and textual approximations of glaciers travel across timespace and generations as landscape descriptions, measurements, maps, models, theories, mimicries, and artworks. When appropriate in number, quality, and kind, the collection in whole or part layer into the corpus of sedimented and emergent geoscience.

Scientific filters that ascertain data's significance change generationally. But if confirmed as reliable, data abstracted from original acts of collecting – from being there, on the ground, influenced as directly as possible by material evidence issued from the sphere of unintended agencies – become part of the conceptual foundation for all that follows, from engineering and law to folklore and art. However inspired or distorted by logistics, luck, imagination, or cultural and scientific milieu, those moments of in-the-field contact link those who are to come, like this ethnographer, to the scientific understanding of water bodies in deep time.

In this chapter, I illuminate the culture of geoscientific data collection that revolves around the glacial origin and post-glacial demise of Lake Agassiz. The once-actual geophysical entity has been kept alive as an intergenerational object of geoscience research, and its name and morphing image have become part of Manitoba's identity.

Unfolding in tandem, the legacies of outbursts signalling ice age end have become wellsprings of Winnipeg's contemporary logistical power and cultural identity.

Lake Agassiz is the subject of a branch of ice age theory inspired by erratics. Humans awaken to the ice age via curiously precise yet serendipitous evidence of glacial action. Giant boulders, pulled out of their settings by inconceivably powerful glaciers pushing across the earth's surface, get dropped off somewhere else far away. I want to begin the story with local farmers and hunters in the Swiss Alps who intuit the agency of boulder-moving glaciers flowing across land (though their tracks through the published literature be wispy). Sometime between late eighteenth and early nineteenth centuries, just before the Victorian era, records show that locals share their knowledge with intellectual adventurers who climb up into Jura mountain valleys from towns in the lower altitudes.[1] The cross-fertilization of local and scientific knowledge transforms the icy mountains into a centre and source of ice age discovery and debate. In an 1828 treatise, a naturalist-geologist names the boulders erratics in recognition of their magnitude and nomadic character.[2] Erratics inspire artists and architects as well as scientists in distinct and collaborative modes of geological meaning-making.[3] Indeed, as Bobbette and Donovan argue (2019, 7) in their introduction to *Political Geology*, when geologists use aesthetic tools to set rocks "into motion," they change how people "see and sense."[4]

As systematic as it is, today's geoscience depends on methodologies of roaming and on logistics that facilitate erratic encounter. The scientific importance of erratics derives from the way that their mineral composition indexes their origins and, too, how their origins contrast with the mineral composition of the sites where humans encounter them. The contrast signals that the elemental relationship of rock and ice is one of travel. This persistence of erratic character in the context of radical spatial movement attracts the attention of natural historians, philosophers, geologists, and artist-collaborators working in overlapping and morphing professions (a precedent for today's transdisciplinarity).

Well-travelled erratics enable human geological actors to map ice flow vectors based on data points of pickups and drop-offs. As the data-corpus grows, ice age theory circulates and inspires new field research in other post-glacial regions. Data on erratics coming from

North America, for example, suggest to early nineteenth-century geologists that, although erratics provoked the emergence of ice age theory in the Jura Mountains of the Alps, the glacial forces that moved them off the north polar region, and the geological phenomenon at stake, were worldwide in scope.[5]

Ice age theory travelled from local shepherds and hunters into geology by following erratic patterns to their source in glacial power. As the nineteenth century proceeds, erratics awaken geology to two interrelated ideas: 1) earth has experienced various ice ages in which glaciers dominate continents; and 2) ice ages are interposed with warmer periods, called interglacials, in which glaciers remain frozen solid only at poles and high altitudes. (Humans inhabit earth's most recent interglacial.) Chapter 2's action begins at the end of the Victorian era with my reading of selections from Upham's (1895) comprehensive work on Lake Agassiz.[6] Geologists have basically accepted the two inter-related ice age ideas by this time. The analytic descriptions of Upham's journey show how horse and wagon fieldwork practices produce glacial theory by enacting the flood bowl that would become the province of Manitoba. Together, descriptions of the unique pattern of emergence of the four main water bodies – Red and Assiniboine Rivers and Lakes Winnipeg and Manitoba – provide readers with a foundation for understanding the flood threats that challenge the twenty-first-century river city and its environs.

Following this aquatic origin story told through a nineteenth-century lens, a contentious mid-twentieth-century conference brings readers to another telling moment in Lake Agassiz studies: aerial photography has just changed the way data are collected and interpreted. A technology that visualizes earth's surface through the eye of a camera in air, rather than the eyes and hands of a person drawing in a notebook, has become available. Aerial photography shifts the parameters of data collection and thus of data interpretation. Some geoscientists are more comfortable than others with the pragmatic notion that not all fiction can, or should be, winnowed out of authoritative fact production (Elson 1967, 85).[7] This chapter's path ultimately leads to the Outburst Quartet, the dramatically speculative and technologically sophisticated twenty-first-century model that builds on, yet transcends, all the previous geoscience on Lake Agassiz (Teller, Leverington, and Mann 2002; Clarke et al. 2004). Erratics unexpectedly roll into and out of the text throughout.

II. DECIPHERING PREHISTORY IN THE 1890S

Above all, though, as Agassiz sees it, the glacier is itself an enormous writing instrument. A world that to others seems stationary, cold, and lifeless is, to the scientist who knows how to read it, full of life, motion, change.

Christoph Irmscher (2013, 72) in *Louis Agassiz*

A. Theory, Measurement, and Mirage in Upham's Prairie Vision

Like the glaciers it melted out of, Lake Agassiz inscribes intricate patterns of flux onto the continental interior. In its 4,000-year heyday, it creates enough traces of its morphing shape that with systematicity and simple tools, late nineteenth-century geologists compose a great North American origin story. To help me tell an ethnographic version of the emergence of one important strand of this origin story, I turn to Warren Upham's 658-page work, *The Glacial Lake Agassiz*, printed by the United States Geological Survey in 1895. My university library has carefully preserved and loaned me an original printing in a cardboard box. A treasure trove of hydrogeological theory and descriptive detail awaits me. These include comprehensive measures, maps and synthesis of variations in the water-influenced patterns encoded in sand, stone, clay, and gravel.

In 1887, after working in Minnesota and the Dakotas and crossing a still somewhat tentative forty-ninth parallel into Manitoba, the scientist-author and his assistant, one Mr Young, travel three to ten miles a day, mostly afoot (Upham 1895, 7–9).[8] The duo travel between sites in horse and wagon covering much of the same terrain in which, 127 years later, I would travel in my Prius while doing fieldwork. This is no coincidence, since both he and I, albeit for different reasons, are drawn to the exact centre of the continent seeking to understand water's dominion in the flood bowl. (Although, like other geologists and natural historians, Upham adheres to a strict division between nature and culture and focuses his scientific gaze on the side of the former. He rarely mentions humans and seems uninterested in cities or geopolitics.) Reading around his mammoth text, I follow his path along ancient beaches and by cliffs undercut by waves as he marks stages in Lake Agassiz's rise or decline or as he examines the composition of an intruding rock layer to figure out its provenance. As Upham writes up his fieldwork, he folds in data and theories of his contemporaries, occasionally reaching out with curiosity to wonder what geologists of the future might find. And indeed, giving him the respect he is due, kindred souls of future geoscience who follow him into Manitoba continue to site his early work (e.g., Elson 1967; Clarke et al. 2004).

As the first Anglo-European geologist to systematically map and measure the contours of a vast ancient lake once lying atop what is now the drainage basin of the Red River and Lake Winnipeg, Upham exercises his professional privilege to name it after Agassiz, the naturalist historian who convinces the scientific world that glaciers – land ice rather than sea ice – could carry those puzzling erratics across the earth's surface.[9] As Upham walks the prairie landscape with Agassiz's published works in mind, he searches for afterimages of the lake's icy waters. When he identifies them, he levels his tools and checks altitude calculations against the rails of the recently built railway.

Day by day, site by site, Upham carefully documents measurements and observations that guide his interpretation of water's elemental powers. Collecting and collating data, he illuminates the broader interactions of ice/water and sediment moving across the piece of earth's surface within which Winnipeg arises. His scientific language can be evocative, as in this description of erratics (136):

> When the ice-sheet was finally melted, its inclosed bowlders [*sic*] were dropped, and they now lie frequently as conspicuous objects on both the lower and higher parts of the land. Scattered here and there in solitude, in an expanse of prairie, or perched on the sides and tops of hills and mountains, *they at first suggest transportation and stranding by icebergs or floe ice.*
>
> [My emphasis in italics]

The reference to "stranding by icebergs or floe ice" may allude to either bible-friendly interpretations that Upham acknowledges but rejects or to the more scientific, but ultimately discarded, geological deluge hypothesis.[10] The brief allusion signals that whether it is the earlier geological deluge or the biblical flood that Upham has in mind, he does not consider either theory significant enough to warrant more direct discussion. For this late nineteenth-century fieldworker, the relationship between Geology and Genesis has become, as historian of earth science Martin Rudwick (2009) might say, an "amicable dissociation."

The ice sheet quote highlights the poetic style of landscape description in science writing. Evocative passages punctuate detailed descriptions of Manitoba's flood bowl formation. They may affectively enhance empirical content and, too, affectively push geoscience toward the sublime. As such, an empiric-poetic hybrid style may inadvertently contribute to the making of (post)colonial empire more powerfully than either would apart. It is, following Lefebvre (1991, 311–12), a style that rests on a projection, a systematic theory of reality, onto underlying space. With the aim of validating ice age theory and the theoretical project of geoscience more generally, Upham names and characterizes the changing shape of Lake Agassiz, the deep time ice/water body that once inhabited the space underlying his feet.

Upham's text resonates with the mid-eighteenth- and nineteenth-century science and natural history expedition writing that Mary Louise Pratt (1992) reinterprets in *Imperial Eyes*. Evocative imagery in his field-based texts, drawings, and maps conform to this genre of exploration and contribute to a form of planetary consciousness that emerges from contact zones in which explorers produce special knowledge for distant audiences (Pratt 1992, 16, 23).[11] Not unlike Alexander von Humboldt's early nineteenth-century reinventions of South America, which, as Pratt (1992, 39, 121) finds, "fuse the specificity of science with the esthetics of the sublime," Upham's work participates in producing global-science authority in an almost innocent anti-conquest fashion (that is, a privileged fashion not *explicitly* involved in "conquest, conversion, territorial appropriation and enslavement"). Upham probably comes to the empiric-poetic rendering of natural forms via Agassiz, who had absorbed and had made more scientific his mentor's, Alexander von Humboldt's, message to his readers to "contemplate the whole of nature as if it were a well-executed painting (Irmscher 2013, 89)."

More explicitly, the empiric-poetic style helps readers to engage aesthetically with geophysical processes, as intended. Upham leads me to understand, for example, how snowflakes fall through time, layering and hardening to form glaciers (as I write in an opening passage of this book). In addition to his careful rendering of the composition and disposition of earth materials moved by ice/water, Upham has a knack for describing physical properties of phenomena that others might be content to not see, to ignore, or to simply note in wonder. For example, in his renderings of different mirage-effects, a surreal architecture of human settlement unsettles the boundary between sky and land and prefigures the cinematic power of the *Summerland* mirage in the 2020 film directed by Jessica Swale:

> A more complex and astonishing effect of mirage is often seen from the somewhat higher land that forms the slopes on either side of the plain. There, in looking across the flat valley a half hour to two hours after sunrise of a hot day following a cool night, the groves and houses, villages and grain elevators, loom up to twice or thrice their true height, and places ordinarily hidden from sight by the earth's curvature are brought into view. Occasionally, too, these objects, as trees and houses, are seen double, being repeated in an inverted position close above their real places, from which they are separated by a very narrow, fog-like belt. In its most perfect development the mirage shows the upper and topsy-turvy portion of the view quite as distinctly as the lower and true portion; and the two are separated, when seen from land about a hundred feet above the plain, by an apparent vertical distance of 75 or 100 feet for objects at a distance of 6 or 8 miles, and 300 to 500 feet if the view is 15 to 20 miles away. Immediately above the inverted images there runs a level false horizon, which rises slightly as the view grows less distinct, until, as it fades and vanishes, the inverted groves, lone trees, church spires, elevators, and houses at last resemble rags and tatters hung along a taut line. The traveler in the Red River Valley is reminded, in the same manner as at sea, that the earth is round. (21–2)

In this evocation of a peculiar phenomenon, Upham likens the prairie to a sea with the power to reveal the earth's roundness. He provides enough detail regarding the position and timing of his observations to allow others, who should wish to, to replicate his experience, thereby laying the groundwork for further empiric-poetic understanding. Even as it erases the Indigenous inhabitants, the planetary consciousness he creates with this mirage works as much through wonder as through empirical observation and analysis.

The mirage description is one of the infrequent moments where evidence of human activity intrudes upon his thoughts. Another textual moment of human consideration occurs when he encounters geographical conditions that limit the large and singular task he has set himself – deciphering the relationship between Lake Agassiz and its ancestor, the Laurentian Ice Sheet. In a "part rolling and hilly woody region," east of the flat prairie of the Red River Valley, he finds (29):

> a difficult district for exploration, as the greater part of it has neither settlement nor roads, excepting those of the scanty population of Ojibway [Anishinaabe] Indians, who maintain themselves chiefly by hunting and fishing ... Their abodes are usually on the shores of lakes and streams, which they navigate in birch-bark canoes; and this is the only practicable means of travel for geological exploration.

Throughout his travels in this region of eastern Manitoba and before that in Minnesota, he confronts miles of tamarack swamps, morasses, and quaking bogs called "muskegs" that can only be crossed when frozen. (Maybe he or Mr Young can't swim or are loathe to lose the wagon's capacity to properly stow instruments and notebooks or to lose the land-railway triangulation method he uses to measure the earth.) Retreating in defeat, his account registers his disappointment. In the same paragraph, Upham's summary subtracts the Ojibway from the landscape, casually enacting *terra nullius*:

> On account of the impracticability of tracing the shores of Lake Agassiz through this wooded *and uninhabited region*, the northeastern limits of this glacial lake, where the shore in its successive stages passed from the land surface to the barrier of the receding ice-sheet, remain undetermined. [My italics]

In his writing, he accords Ojibway people moving lightly through muskegs in birch-bark canoes neither significance nor respect.[12] And unlike fur traders and immigrant-settlers, who depend on Indigenous knowledge for navigating interior waters, Upham finds no reason to attend to their ways of life or to attempt to benefit from their knowledge of landscape features and hydrogeology, much less spiritual wisdom.[13] Nor perhaps do most of his readers expect him to. And yet, I can't help but consider what insights, cartographies, and telling configurations of ice/water and rock the Ojibway might have shared with him had he asked and had they been willing.[14] Perhaps the lack of connection was for the best, given the risk of viral transmission from outsiders to Indigenous peoples with no history of smallpox or influenza and thus no antibodies to protect them.

On the matter of logistical limits on research scope: Even without the problems of forest and swamp cover blocking his path during the warmer seasons that allow for fieldwork, a lack of transport infrastructure elsewhere also stymies the fulfillment of his intellectual desire. On the open prairie of western Manitoba, his horse and buggy bring him and Mr Young only so far. In Riding Mountain, he can study the action of the western border of Lake Agassiz at its highest stage where its waves lap up against the Manitoba Escarpment. There the road ends, leaving him with significant but, nevertheless, partial assessment of Lake Agassiz's fluctuating boundaries. And too, it may be a lack of infrastructure for measuring variegated elevations against a uniform standard, since the railway is also probably as absent as the roads. But that, in fact, is the nature of science. No matter the sophistication of transport infrastructure and technologies of visualization, the production of knowledge always meets limits. On the one hand, it is always partial, bounded by practicalities and blind spots. On the other hand, science is very good at delineating the shape of uncertainty (see chapters 7 and 8).

I am pulled toward Upham's nineteenth-century landscapes of inquiry, finding inspiration in his walkable visualizations of the last ice age. At the same time, I am somewhat astonished at how oblivious he is of geopolitics, the human struggle in and with the same kind of landscapes he studies. An American geologist, he has no stake in Canadian territory other than as an object and source of deep time knowledge to be shared with contemporary and future geologists. Indeed, because he calls Canada "British America" in his text, it seems that he does not realize that Canadians established independence from Britain less than thirty years before his arrival nor that when the new national capital orders a provincial survey in 1869, it triggers the Red River Rebellion led by Métis leader Louis Riel, lending Manitoba status as the only Canadian province founded by an act of armed resistance (Kives and Scott 2013, 28). Indeed, neither the politics of mapping nor Winnipeg, arguably an important city close to his endeavours and once under Lake Agassiz, fall directly within Upham's ambit.[15]

The railway provides a contrast between Upham's geological focus and his geopolitical context. For him, the railway provides a "network of altitude control points,"[16] a technology of vision and measurement.

For Winnipeg inhabitants, in contrast, the railway signifies a dramatic turn of fortune. When it arrives in 1881, it triggers a thirty-three-year economic boom (Kives and Scott 2013, 38–40).[17] In its first ten years, the period overlapping with Upham's fieldwork, the railway brings thousands of European immigrants to the lands of First Nations and Métis peoples, where they mix with those who stay and survive and with the French and British settlers already there, tripling the urban population to 27,000; by 1911, the city boasts 150,000. Wealth flows into the city with the goods that moved through from the eastern seaboard to the still "Wild West." At the height of the boom, Winnipeg, like New Orleans, always "a city of geographical necessity" (Knowles, 2014), became the fastest-growing city in North America, outpacing Chicago as a distribution hub.[18]

The point here is that, insofar as I can tell from his book, none of the dramatic human history unfolding in Manitoba while Upham's earthwork is in progress matters to him. The province provides an infrastructure and a natural laboratory for the objective analysis of nonhuman ice age events that can be compared to analogous events on different continents. Over time, the mapping, analysis, and publication of these events contribute to the construction of a global theory in earth science. Upham's purpose is to circulate this knowledge through scientific networks well beyond the valley in which he walks. When he finishes, he goes home. Possible allusions to the science-religion debates of nineteenth-century Anglo-European culture aside, this mode of being and knowledge-making in the landscape – a geoscientific mode – contrasts dramatically with the modalities lived by settlers and Indigenous people who are his contemporaries and who inhabit the flood bowl that links him to its ancestor, Lake Agassiz, the object he names and studies.[19]

In any case, despite or because of the limitations Upham finds or imposes upon himself, the myriad elemental forms, dispositions, and effects of ice/water come alive in his text as a composite, shape-shifting actor that assumes dominion over the planetary surface through its interactions with wind, soil, and stone. I mine Upham for work that pertains more specifically to the inter-linked water bodies most important to Winnipeg's modern flood control system – the Assiniboine River and Lake Manitoba and the Red River and Lake Winnipeg. The next section focuses on his nineteenth-century version, highlighting their emergence and infrastructural relevance.

B. Upham's Hydrogeological Findings

Glaciers, rivers, and lakes write earth's stories by sorting, transporting, and releasing sediment.[20] Through their different expressions of order and chaos, glacial inscriptions convey earth's dynamic and complex transformations over time. By studying the kinds and dispositions of materials and locales, Upham discerns the prehistories of ice/water flows that created the topographies upon which he and his assistant travel. Mapping ice/water flows, he scales up his data on sediment transport to trace the broad outlines of Lake Agassiz. The following paragraphs sketch the prehistory of water bodies most relevant to Manitoban flood control, only a little piece of the descriptive terrain Upham covers.

The flood bowl takes form before the last ice age. Its network of interlacing river basins in the continent's centre forms a braided trough that drains northward toward the pole. When glaciers of the last ice age build and advance southward into the river basins, they meet bedrock. As the glacial lobes and rivers entangle with each other, the bedrock guides and resists glacial movement. For its part, the glaciers break off pieces of bedrock and plow up surface materials. The bedrock and surface materials change geological identity in the force of encounter. Once travelling, they become "drift." Entrained within a moving glacier, the drift then writes upon the bedrock it scrapes over, leaving fine scratches and marks called striae. In other words, the glacier makes bedrock into writing instruments *and* surfaces for writing upon. Upham (1895, 108) finds striae in the Red River Valley.[21] "Only one cause is known which can produce markings like these," he writes, "and this is the rasping of stones and bowlders [sic] frozen in the bottom of a moving mass of ice accumulated upon the land in a solid sheet of great extent and depth."

Releasing the drift as they melt and retreat, glaciers add another layer of informative chaos. Once separated from its ice transport, the drift changes geological identities again. No longer drift, in technical language, the material becomes "till." Till is sediment that can be transported anew by water, wind, or gravitational force. Each element or force that picks up and moves the till encodes its distinctive mark through the interactive processes of its subsequent release. When till falls back to earth's surface, it may reveal the nature and path of its transporting element or force if a perusing geologist finds it (Upham

1895, 35, 108).[22] Till (and here I stray from technical language) is like an assortment of teeny tiny erratics that arise to visibility only in collective masses.

Consider, for example, the action of rain in the formation of Lake Agassiz: rainstorms wash over dissolving ice, intensifying floodwaters. Ice/water rivers spill away from waning glacial giants with extraordinary force. Rushing across the drainage basin toward the North Pole, the ice/water clears a path, washing away "detritus" (a word suggesting that some sedimentary formations are not as useful to geologists as drift or till). How much sediment rainwater moves varies over the time of Lake Agassiz's existence. And estimates of rain's effects change with advances in geological measurement (Upham 1895, 63). In any case, when transported in the currents of rivers or into the relative stillness of lakes, sediments may sift out more or less evenly, forming stratified layers available for study.

Eventually, riverine meltwaters begin to dominate the surface. But even in retreat, the ice sheet holds its power, its conatus, or will to persevere. When flooding meltwaters move north with gravity's downward force through riverine networks toward Hudson Bay, they cannot flow out. They are blocked by the northernmost edge of the retreating ice sheet. The ice dam captures the flow of its own meltwater as if refusing the complete loss of coherence entailed in oceanic dispersion. Trapped, the rivers continue to spill out of different parts of the retreating glaciers and come together to form lakes. The biggest of these is Lake Agassiz, which edges right up against the ice dam and fills up about one-third of the flood bowl.

Water rushing northward from the continental divide (in present-day North Dakota) digs a shallow channel under Lake Agassiz that eventually becomes the Red River. Some rivers, like the Assiniboine, arrive at Lake Agassiz with their own preglacial histories and sustain their identity after Lake Agassiz is gone. These past relationships and identities persist in Winnipeg's present, shaping the possibilities and limits of flooding and flood control.

The Red and Assiniboine Rivers are lost in the vastness of Lake Agassiz. They meet each other as discrete water bodies only after Lake Agassiz is gone. In other words, the birth of their confluence is predicated on the death of the lake, yet in extreme flood, their entwined existence keeps the ghost lake alive. The destiny of Red and Assiniboine Rivers unfolds in relationship to two large, long lakes, Manitoba and Winnipeg.

1. *From the South, The Red River and Lake Winnipeg:* When the largest ice barrier confining glacial meltwaters in Lake Agassiz gives way, it allows the waters to burst out into Hudson Bay. Afterwards, an ancestral Red River continues to wend its way north through the trough connecting the once southern limit of Lake Agassiz with the once shoreline of Hudson Bay (smaller and closer to Winnipeg than it is today). A strand of small lakes expand off the ancestral Red, eventually coalescing into the long, shallow form of Lake Winnipeg. Over the millennia, Lake Winnipeg gets bigger and bigger, engulfing more of the Red River within its basin and leaving the Red River shorter and shorter. Eventually, the waters reorganize to become the interconnected bodies whose shape appears on regional maps today. The Red River and the long and shallow Lake Winnipeg are divided only by the fuzzy boundaries of Netley Marsh.[23]

This geoscience knowledge is the basis for my ethnographic interpretation of the Red River's contemporary character – in particular, how the Red is haunted and empowered by a reoccurring capacity to spread out into a lake (or sea). This interpretation is well supported. For example, historian J.M. Bumsted (1997, 8) explains how the physical features of the Red River Valley are conducive to flooding:

> Beneath four to 60 feet of clay – topped by some of the deepest and richest soil in North America – exists a glacial till of rock, clay, sand, gravel, and boulders. The slope of the Red is very gradual, averaging six inches per mile from Wahpeton, North Dakota, to Lake Winnipeg and only three inches per mile in its Canadian section. Because of this gentle slope, the Red is not a very swift-flowing river. It has not managed to cut deep channels. When at flood time large quantities of water flow through the Red, they cannot be contained in the relatively shallow channels. The water spreads quickly over the surrounding areas, halted only by uplands some distance east and west of the river.

Although fairly stable most of the year, the Red River enacts a seasonal impulse that mimics its glacial forebears. In an already soggy prairieland, a spring ice melt can turn the Red River into a freshwater sea that drowns Winnipeg and its surround. This seasonal recall of Manitoba's glacial origins is fairly predictable. Engineers have successfully designed and built surveillance and infrastructural systems to protect against the Red's spring floods (see chapter 3).

2. *From the West, the Assiniboine River, the Delta, and Lake Manitoba:* The Assiniboine River might be much smaller than the Red River, but it is older and more complicated and drains a much larger area. The Assiniboine is older even than the glaciers of the last ice age. Rivers excavated the irregular surface of its basin before the Laurentian Ice Sheet moved down from the North Pole. And each of its major tributaries such as the Qu'Appelle and the Souris has its own distinctive history (Upham 1895, 59). When the expanding ice sheet does move into the Assiniboine basin, it releases a thick, fairly uniform layer of drift across the landscape. The deposited sediment, the till, preserves the broad outlines of the Assiniboine basin's existing contour. And so, the water that courses across the surface today moves along topographical variations sculpted by preglacial, glacial, and postglacial ice/water, including an unusual feature – a river delta in the interior of a continent.

When the Assiniboine River flows into Lake Agassiz at its highest level (when lake waves lap against the Manitoba Escarpment), it essentially encounters a freshwater sea. In that space of encounter, the seventy-five miles between the present cities of Brandon and Portage la Prairie, the Assiniboine forms a giant delta of sand and gravel about 2,000 square miles in extent and about 50 feet deep (Upham 1895, 189). Upham estimates that a third to a half of the sediment comes from glacial drift, which the Assiniboine carries in from a glacier melting in its upper basin; some drift is still entrained in blocks of ice that melt and release sediment only after they reach the lake. Another area is covered with alluvium that the Assiniboine produces by ordinary river erosion. As the river rushes through the delta of its co-creation, it also picks up fine grains of silt and clay that eventually come to rest in layers on the lake bottom. Different modalities of sediment transport and deposition leave patterns that can be deciphered. By comparing unstratified and stratified sediments – those that chaotically tumble and those that calmly sift – Upham begins to work out the origin and shape of the river basin, contributing to an intergenerational geoscience project (see more below).

The Assiniboine River outlives Lake Agassiz and enters into a relationship with several bodies of ice/water that take its place – most importantly, Lake Manitoba, one of the largest remnants of Lake Agassiz. As it begins to journey out of the west, the Assiniboine is first confined to deep valleys and channels; these then flatten out into wider and shallower channels as the Assiniboine flows over the delta formed by its ancestral encounter with Lake Agassiz. As it flows past Portage la Prairie, south of Lake Manitoba, a geological dialogue between river and lake is sustained by a low networked topography that has the potential to flow bidirectionally, south to north and north to south.

The Assiniboine River and Lake Manitoba have no direct connection like the Red River and Lake Winnipeg. But the river flows right past the bottom (southern end) of the lake. And in the space between them, the watershed is so low that when the river overflows its banks, it sends floodwaters into the lake. And too, in short periods when Lake Manitoba is at its highest level, it can become a tributary to the Assiniboine. This geological dialogue lays the foundation for the Portage Diversion, an engineering structure designed for flood control that mimics, albeit in a unidirectional and controlled fashion, the movement of water between river and lake. And as it happens, between Portage la Prairie (where the Portage Diversion is sited) and Winnipeg, fifty-three miles of road surrounded by farms and suburbs, the Assiniboine's channel is extremely shallow, and the river overflows easily (Upham 1895, 380; Bumsted 1997, 8). This part of the floodplain is a trouble spot in provincial flood control.

In some cases, geophysics can provoke international geopolitics (Dodds and Nuttal 2016, 154) or within sovereign territory, provincial geopolitics. When floods pour across the western plains into the Assiniboine basin, engineers can split the Assiniboine River flow at the Portage Diversion[24] to send excess water northward, thereby preventing it from heading down toward The Forks in central Winnipeg. Because this operation protects the city by displacing flood hazard on those to the north and west of the city, the geological dialogue between river and lake gives rise to governing policies, political and legal procedures, and arguments. In any given crisis, water engineers and politicians can decide to send floodwaters to Lake Manitoba to protect the city. If, because of this decision, they trigger disaster for people who inhabit the lake's edge, then justice requires compensation for affected inhabitants. As briefly introduced in chapter 1, legal compensation entails a calculative process in which "artificial flooding," caused by flood control operational decisions, is separable from "natural flooding," caused by a combination of elemental forces that, in the moment that disaster strikes, are beyond human control. The natural is thus defined as the flooding patterns that *would have been* if modern infrastructure had not intervened in flood events (see chapters 4, 5, 7). Safe to say, this is a complicated spot.

In short, the Red River – a remnant of Lake Agassiz – and the Assiniboine River – which precedes and outlasts Lake Agassiz – are melted configurations (re)shaped by the Laurentian Ice Sheet that pass through very different phenomenological phases. Despite different glacial histories, however, their gumbo-lined confluence intensifies both of their geo-cultural powers as they pass through, or are prevented from passing through, Winnipeg. Together, the rivers organize all the infrastructure-enabled movement in the city. Although the Red and Assiniboine Rivers have lost the central transport role they played in the time of canoes and York boats, provincial roads and railways continue to weave around their meandering shapes, differently affecting patterns of neighbourhood mobility. And their confluence at The Forks continues to provide a meeting point for inhabitants and travellers.

III. GEOSCIENCE DATA INTERPRETATION

Isobase: a line (either imaginary or on a map) connecting points on the earth which have undergone equal amounts of uplift (or more rarely depression) over a period of geological time.

Oxford English Dictionary

Lake Agassiz continues to be an object of geoscientific study as a theoretical project with practical implications. From an ethnographic perspective, the lake is at most a once-real place, a theoretical, partially definable place arising from the sphere of the ice sheet's unintended agencies. Lake Agassiz is an empirical earth story about a complex process that takes place over four millennia and leaves inscriptions on the landscape that continue to resonate across deep time.[25] It is a 200,000-square-mile area over which ice/water spills out and pools from multiple directions, submerging different surfaces at different times while picking up and releasing sediment, a non-linear event at the end of the last ice age. The glaciers melt, retract toward the pole, and advance again. All that glacial back-and-forth destroys the data-story that the lake's meltwater waves tell, fragmenting the evidence chains available for future geoscientists and compelling interpolation in the gaps of the known (Elson 1967, 94).

The data-filled stories come into being through technologies of measurement and visualization. They start with field geologists like Upham who walk and measure land contours at different elevations, then plot the line connecting the points at highest elevation to create isobases. This graphic method for elevation has been used at least since 1892,[26] although as Ferrari, Pasqual, and Bagnato (2019, 63) point out in their study of the cartographic imagination of Italy's Alpine limits, "the idea that landscapes could be a series of numbers goes back to the early 1800s work of Alexander von Humboldt."[27] By mapping a series of isobases, geoscientists can infer the highs and lows of Manitoba's vanished (and vanishing) ice/water formations.[28] As in any map-making process, the spacing between the points and hence the accuracy of the data collection and the eventual textual, graphic, or arithmetic representation of the earth contour must accommodate the situation on the ground.

But geoscientific construction of Lake Agassiz is an aesthetic as well as technical process. There is no absolute right number of dots per inch. Lines in the earth and representing the earth are made of contrapuntal folding and unfolding of straight lines and curves. Evocative landscape drawings, paintings, maps, models are theory-machines. And the continuities of continent-crossing relational standards like rail lines that happen to be built for moving goods also enable geoscientists to pull out a nuanced spatial presence like Lake Agassiz out of deep time (like rabbits from magicians' hats?). In Upham's work, the rails (and the beds upon which they sit) serve as a fixed baseline, a measurement infrastructure for the production of scientific knowledge. The railway lines are useful to Upham as orderly arrangements of wood and iron embedded in a variegated prairie surface. Their arrangements offer a ready-made quantitative index of land level in relation to sea level that is backed by official consensus. It converts place-based apprehensions to global data points. Yet railway lines are not simply neutral reference devices. They are also amalgams of political and economic decisions by surveyors, investors, industrialists, lawyers – expressions of capitalist logic and logistical power. All these activities, these territorial imperatives, precede Upham's arrival, who imports the data produced by this calculating machine into his work. From field, to monograph, to archive, data get layered into the heart of the intergenerational geoscience corpus. In the sphere of unintended agencies, then, we come to understand Lake Agassiz through an empirical but nevertheless distorted lens. The entwined destinies of ice/water and industry rendered in maps and models is a theme that shall recur in this ethnography.[29]

A. Spaces between Points:
A Conference in Winnipeg about Lake Agassiz, 4–5 November 1966

As J.A. Elson (1967, 37–8) explains to a group of geoscientists and members of the public, the boundaries of Lake Agassiz vary considerably. In some parts, well-developed beach ridges and wave-cut scarps and terraces can be traced with ease. In others – for example, where the boundary is an archipelago of drift-covered bedrock islands that inhibit wave action – boundaries are more difficult to ascertain. (It was at just such difficult spots, like the muskegs, where Upham's descriptions met their limits.) But suddenly, the geoscientists find themselves confronted with a dissident who does not believe in the validity of their interpolations. Unusual for proceedings of an academic conference, Elson steps outside the frame of his straightforward presentation of findings to elucidate the problem at issue for his audience (85):

> Professor W. O. Kupsch severely criticized all work on relict strandlines and condemned the work done on Lake Agassiz ... He implied that no strandlines were traced continuously, but that they were merely interpolated between widely spaced sets of measurements, the relationship of which to the water planes was unknown. His further comments on the errors in drawing isobases and on the interpretation of their tectonic significance was indeed well taken. However, the impression he created for non-geologists of his audience was that sloping water planes are imaginary and not valid evidence for the interpretation of geologic history. This view would be contested strongly by a majority of geologists.

Elson's remarks come seventy years after Upham published his tome. His generation does basic science with practical relevance. They have access to newer technologies of data collection and visualization. Aerial photography has extended the capabilities of these 1960s geoscientists. Geoscientists doing research at this time are able to incorporate measurements of parts of the basin that had been unreachable before. Yet sometimes, even with the added benefit of the new technology, interpolation still must be relied upon to determine where a presumed water plane intersects with topography. In his conclusion, Elson (94) addresses Kupsch's critique again, pointing out the inevitability of weak links, which in any case are mainly details in a chain of evidence characterizing a glacial phenomenon that destroys some traces even as it inscribes others. The "hypothetical history of the lake" presented at the conference, he argues:

> should be regarded by non-geologists as an illustration of how evidence can be fitted together to establish a continuous story of Lake Agassiz. This four-episode history is a substantial advance ... but it still contains much speculation and undoubtedly some fiction.

In Elson's introduction, the use of aerial photography is presented as a technological boon that simply extends the scope of walking-the-surface data collection. But perhaps the new *vision tool* introduces new interpretive doubts about what the spaces between data points mean. As it turns out, except for Kupsch, the confidence that Elson and his colleagues have in the four-episode history of Lake Agassiz is warranted (see more below). But one conversation that does not make it into the proceedings and might not have occurred is worth contemplating, i.e., the role that technology plays in the negotiation of scientific uncertainty. I find it in the published archive.

As Matthew Dyce (2013) argues, the post–World War I introduction of aerial photography into civilian surveying and mapping endeavours does more than simply extend the scope of data collection. Photographs taken from the sky flatten the ground, changing the perspective on earth forms – their shape, patterns of light and shadow, inter-relationships. Selected and framed in straight-line series, three-dimensional earth features are projected on to two-dimensional Cartesian planes on paper photographs. Airplanes, cameras, pilots, and data interpreters who work on coordinated aspects of disembodied (or differently embodied) image-making take the place of authoritative scientists and cartographers in the field whose data entail sensory as well as technical transcoding of topography. The basis for establishing objective truth changes, a cultural shift that begins to prepare the public to interpret life through digital image flows even as it extends the scope of science.

Whether science is assembled by /horse + wagon + railway + field scientist/ or /pilot + airplane + camera + scientist + office computer/, transforming data into theory depends on the active interpolation of that which is believed to be the shape of empirical reality. A leap of faith, perhaps, a confidence that hydrogeological processes will follow an orderly path in defined situations (like water, when no obstacle is present, flowing downhill with gravity), and an aesthetic sense, a bracketing off of the sphere of unintentional agencies. Call it what you will – fiction, inference, imagination – scientists must be comfortable with sustaining the tension between the empirical and the interpretive.[30] Technology, politics, and the erratic earth itself shape how and when to set the balance as geoscientists convert the glaciers' writing and melting into data and public knowledge.[31]

IV. THE OUTBURST QUARTET AND TRANSNATIONAL TRADE ROUTES: ENTWINED GLOBAL IMPULSES[32]

A. Whoosh! X 4, Deep Time Climate Change

My retelling of this origin story is based on Teller, Leverington, and Mann (2002) and citations therein: Over a century after Upham, still working on Manitoban events, geoscientists collectively introduce a radically different kind of imaginative leap into this origin story. They utilize the most sophisticated technological advances in visualizing Quaternary events. Together with quantitative modelling techniques that extend the range and nuance of sensible measurement, geoscientists capture the glacial undoing that forcefully moves abundant fresh water from continental interior to ocean surrounds. These events, they hypothesize with confidence, probably change the course of the oceans' thermohaline[33] currents into the very same routes that sea captains steer their ships upon today.

Lake Agassiz's Four-Act Outburst: By the time the ice dam blocking Hudson Bay collapsed, triggering the largest outburst in the last ice age, Lake Agassiz had merged with glacial Lake Ojibway, to form a superlake with a surface area of about 841,000 km^2, more than twice that of the Caspian Sea, the modern world's largest lake, and larger than modern-day Hudson Bay. Lake-Agassiz-Ojibway is the only lake in North America to have abruptly released huge volumes of stored water (883–5). Outbursts cut through what are now the Mississippi River Valley, St Lawrence River Valley, Mackenzie River Valley, and Hudson Strait. Following the initial outbursts, baseline flow would resume along the new routes to the ocean. Meltwater from these outbursts join the oceans in a very short time period, "in critical locations and at optimal stages in the evolution of ocean circulation ..." Combined with longer-term fluxes of Lake Agassiz, the outbursts "probably played an integral role in global ocean and climate history during the last deglaciation ... *pushing the system over a threshold* (883)." [my paraphrase and italics]

The four large arrows in Teller, Leverington, and Mann's (2002, 880) geoscience article indicate probable routes of freshwater outbursts to the oceans from glacial Lake Agassiz (see figure 2.1). The authors put forth the theory that each episode in which large volumes of lake water suddenly burst out of the confining but collapsing walls of the Laurentian Ice Sheet may alter the circulation patterns of ocean currents and trigger a series of widespread climate changes. In the map-image, Winnipeg, though unmarked, is centrally located in what was once Lake Agassiz, well within the source terrain from which the global impulse projects itself toward the oceans. The artifact collapses the hypothetical time frame of events that unfold over several millennia at the end of the last ice age. It effectively projects the viewers' mind into the condensed past of deep time when the ghostly map lines of sovereign nation-states are not even a glint in the planet's eye.

Teller, Leverington, and Mann theorize that any one of four episodes in which large volumes of lake water suddenly burst out of confining but collapsing walls of the Laurentian Ice Sheet may have altered circulation patterns of ocean currents and triggered a series of widespread climate changes. Ice/water's logistical power at continental scale breaks into planetary scale when it pushes the earth's system over a threshold. Crossing that threshold catapults relations between fresh water circulating in glaciers and rivers and saltwater in oceans into the Holocene configuration that humans find when they arrive and settle in North America.

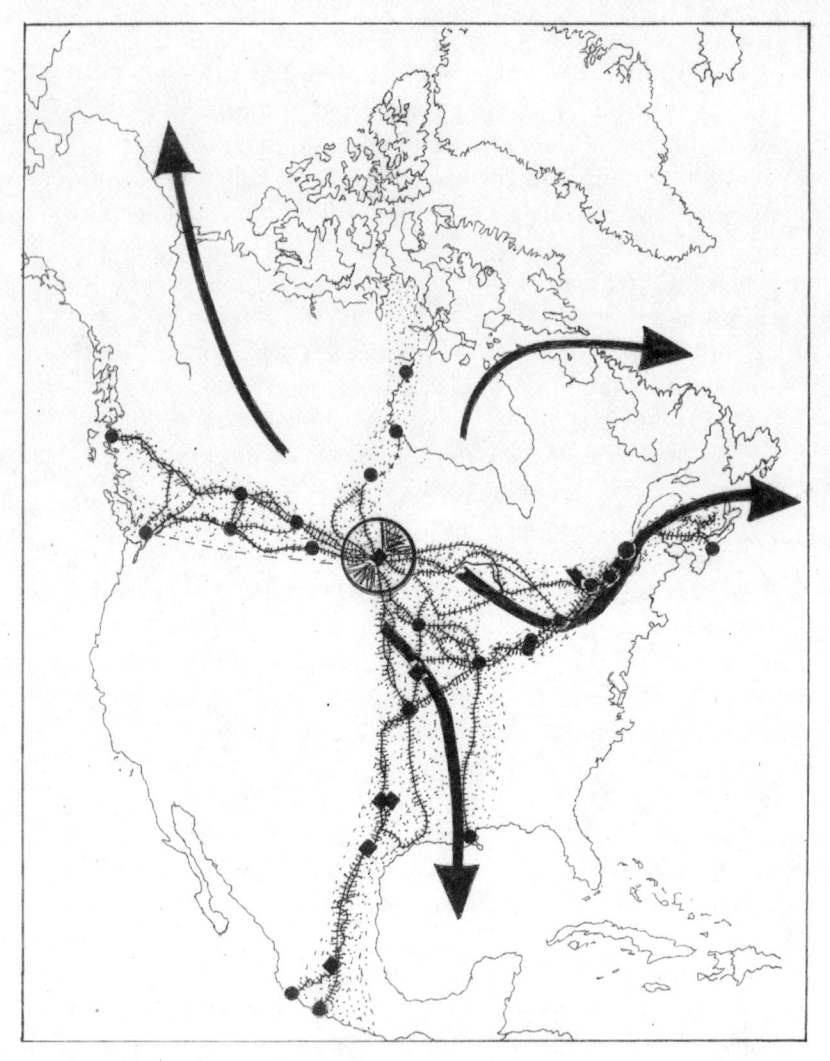

2.1 The Double Artifact: A) Geoscience model of tentacular meltwater outbursts (four arrows) at ice age end (Teller, Leverington, and Mann 2002, 880); and B) CentrePort's map projection placing Winnipeg in the centre of river-city-railway routes linking Arctic and Gulf of Mexico ports. In the background, ghostly contours of NAFTA countries orient perception.

The authors' quantitative model and accompanying narrative briefly sketched here characterizes unfolding events in geological timespace. But the narrative contrasts with their map-image, which collapses into a unified and flattened form the hypothetical events that unfold over several millennia. As a cultural artifact, the map-image effectively projects the viewers' mind into the condensed past of deep time. In addition to condensing time, the map-image contains a cartographic sleight of hand. The ghostly map lines of sovereign nation-states penned lightly into the background insert a geopolitics that is, in fact, not even a glint in the planet's eye.

In comparison to the sophisticated twenty-first-century tools used to model outburst routes, Upham's nineteenth-century strandline elevations and even Elson's aerial photographs are primitive. Yet for more than a hundred years, clear technical traditions and persistent theoretical questions about glacial processes unite the geoscientists.

B. Accidental Experiment and the Double Artifact[34]

While studying Teller et al.'s (2002, 880) twenty-first century image-map of Lake Agassiz outbursts published in *Quaternary Science Review*, I am surprised by striking similarities with an image-map from the webpage of Canada's Centre for Global Trade (Winnipeg's CentrePort).[35] I look at the two heretofore unrelated artifacts. Each depicts a global impulse that connects the geographic centre of North America with four large water bodies (Gulf of Mexico, Arctic Ocean, North Atlantic Ocean, and Hudson Bay). In each, the global impulse is rendered in the form of powerful and expansive tentacles projecting outward from Winnipeg and its environs. The tentacles emerging from page and screen can be located geographically in relation to the faint outlines of sovereign territories and ice/water bodies behind them.[36] Drawn with the thinnest of pens, the continental and island coastlines, the divisions between countries (and in one, states and provinces) appear clear yet ghostly beneath the dramas of spectacular global expansion. The repetition of framing is odd, given that happenstance brings about the artifacts' relationship. The artist-cartographers use almost the exact same base map. So similar are they, that with a little adjustment of scale on the photocopier, I easily combine the images into one with minimal distortion of the content or frame of either. The experiment inspires a visual analysis of intertwining and emanating flows of water, steel, and knowledge (see figure 2.1). Taking each visual representation in turn:

C. Steel Routes of Transnational Capitalism

In contrast, B, the second artifact (now embedded in A, the first) projects the viewer's mind into a potential future when CentrePort routinely brings places like Murmansk, Russia, in the Arctic Circle, together with New Orleans through Hudson Bay and Winnipeg. The tri-modal transport infrastructure (railways, trucks, and airplanes globalized by fossil fuels and shipping) proposes new connections and logistical spaces between existing markets. In the image-map, Manitoba's Arctic tentacle, absent railway lines, remains sketchy. Nevertheless, the optimistic text elaborates on the developing connections among new and older commodity transport infrastructures. The logistics of future success relies on the assumption that shipping will shift northward as twenty-first-century climate change melts the sea ice. Although technically in the geographical centre of North America, in settler practice Winnipeg tends to be more of an inland outpost on the northernmost edge of agriculture and urban life. Among some Indigenous people, however, The Forks in central Winnipeg is known as a powerful place in "the heart of Turtle Island," i.e., of North America (Leon and Nadeau 2017, 119), suggesting that their understanding is remarkably consistent with contemporary urban aspirations and the latest geoscientific modelling.

With climate change, CentrePort, like other ports in ice-prone places, may become more central to global logistics, as city boosters are keen to point out. In this imagined future, climate change is simply an opportunity to ramify transport technologies and diminish the importance of all the borders marking sovereign territories, including the divisions among Canada, the US, and Mexico. Winnipeg was sitting at the northernmost end of the "NAFTA" highway connecting these nations; since 2018, port boosters are going with "Mid-Continent Trade Corridor" (see figure 2.1).[37] In the image-map, the backbone of the future (in the form of tentacles) is still tied to steel railway lines of the sort that propelled the expansion of nineteenth-century frontiers and served Upham as a measurement framework. All living beings, their habitats and evolutionary and revolutionary histories are even more invisible than nation-states.

There is something significant in the trick of global scale that renders space inhuman in these artifacts from geoscience and trade. In each, the conceptual plane formed by the pale outlines of sovereign territories set off the viewers' sense of sheer magnitude and irrepressible directionality of the tentacles projecting across abstract land-sea space. The renderings of global impulses need the barest reference to the world we know so that viewers can orient and sense proportion. This maneuver makes the artifacts effective visual technologies. I realize that by combining them, I read things into them that their creators do not intend. For me, the coincidence in the visual formation of global impulses provides sufficient justification for the ethnographic experiment.

Together, the map-images (a double artifact) hold the codes of an empirically grounded imaginary that stretches from ice age deep time to transnational capitalist futures. Combined, the doubled artifact hints at the "hyperreality" of the originals, the difficulty of separating the actual from the hypothetical when studying an ancient shape-shifting entity like Lake Agassiz or when investing in the multi-modal future of CentrePort (Baudrillard 1983, 2). The sense of hyperreality that arises from the coincident imagery creates doubt. Given the coincidences involved in their manufacture, do either of the single map-images represent something real? Do meltwater outbursts really create four smaller current and climate shifts within the larger overarching climate change that we think of as the end of the last ice age? Will climate change in our epoch unfold such that Winnipeg could become the viable inland port projected on the website gateway diagram? What is behind the visual alliance of faded sovereignties?

Like Kupsch's critique of the 1960s isobases, doubt may insert itself and undermine the authority linking representation to referent, image to object or subject. But Kupsch's doubt does not undermine the use of visual artifacts in geoscience more generally. And even at the 1966 conference, Elson's pragmatic intervention effectively wipes away the doubt Kupsch raises. (The dialogue raises a straightforward methodological question: how many data points are needed to establish the identity of a strandline?) In contrast, hyperreality provoked by experimental juxtaposition of these two visual artifacts inserts a vague sense of uncertainty, anxiety even, regarding the reliability or capability of geovisual representations of the real. Along with important ideas, each artifact may accidentally convey the traces of a suspect ideology that entwines and conflates geo-politics (of rivers and earth) with conventional nation-state geopolitics (of multimodal port logistics). In any case, as the ethnography moves back into the infrastructural workings of the city, macro-scale abstractions break down. Traces of the global appear and disappear in Winnipeg's present urban landscape, and with them, the boundaries between empirical and interpretative knowledge glitter with the sun and grey out with the clouds.

D. Inside Outburst A, Kulchyski and the Deh Cho

I return to the sense of wonder evoked by Upham's careful descriptive analysis of a prairie by turning to *Like the Sound of a Drum*, Peter Kulchyski's ethnographic narrative (2005, 119–22). This book about the Dene and Inuit unfolds within the space broadly defined by Lake Agassiz outbursts A and B, respectively.

Sacred Space in Fort Simpson (Outburst to the Arctic): Returning to the tentacles in the double artifact, note that where Lake Agassiz's freshwaters push east through the St Lawrence Seaway to the North Atlantic Ocean and south through the Mississippi Valley to the Gulf of Mexico, the outburst-tentacles are surrounded by entwining railway lines and urban markets. In contrast, the contemporary human presence in northern Canada is too light to register in this graphic. Outbursts pushing northwest through the Mackenzie River Valley to the Arctic Ocean and northeast through Hudson Bay to the North Atlantic boast only one tentative railway line midway between them (connecting Winnipeg to Nunavut). Yet this vast still-cold space of ice age flows is home to many First Nation and Inuit communities. Their narratives for visualizing the power of ice/water contribute an important dimension to the geo-politics of global impulses.

By reversing the background-foreground relationship in the doubled artifact, I flip the global tentacles into the background and zoom the narrative into the space that appears blank. This maneuver aims to locate the global impulses in a little-known everyday reality in an ethnography that emerges from the far north. Peter Kulchyski, a Manitoban Native Studies scholar writing about the Dene, is my guide. Unsurprisingly, his authoritative style contrasts markedly with the calculation-based image-maps from geoscience and trade. Along with communicating the cultural politics of people and place, Kulchyski cultivates a steadfast awareness that the ethnographic narrative is an artifact created through dialogue and waterscape. I borrow fragments from it to compose a small slice of placetime, to sketch in and qualify the blank space beneath the graphic tentacle connecting Winnipeg to the Arctic.

Fort Simpson, one of many places Kulchyski visited, is situated at the juncture of two mighty rivers, the Deh Cho (Mackenzie) and Naechag'ah (Liard). A traditional gathering place (as tourists are told), the Dene understand the power of the confluence as both material and spiritual.[38] Yet upon seeing it for the first time, the place lacked drama for Kulchyski:

> two big, sluggish, rivers coming together: each so wide that the power is dispersed, too big to grasp. On Nahedeh, a much smaller river, you can feel the pulse that is its power, you cannot forget it. Deh Cho's pulse is so huge that it's muffled, like thunder in the distance.

As if trying to crack the code, Kulchyski is drawn again and again to the point that overlooks the confluence. After many experiences sitting and talking, walking and running along the banks, he finds that "some of the power of the place, of its magnificence, pulsed in me (122)." In this place that brings rivers and people together, the collaboration between this non-Native ethnographer and the Dene-Métis of Fort Gibson leads to an essential yet rarely articulated dimension of the human–water relationship:

> The moral topography here speaks simultaneously to coming together and being swept along; it speaks to underlying forces and currents of unimaginable strength and to the surface that belies those currents; it speaks to cycles and repetitions; it speaks to the everyday banality and to the moment of extraordinary explosive, irruptive power.

It is only through fieldwork that Kulchyski can work through the banality of broad surfaces and repeating cycles to discover compulsive tensions – the coming together and being swept along, the deep forceful currents under surface calm – that inform the cultural meaning of mighty northern rivers in their everydayness. And too, there is the waiting for dramas such as the rites of spring. In the time called "breakup ... enormous chunks of ice crash and thunder and split in the frenetic, explosive tearing apart that leads to renewal" (120). As the rivers re-enact the geological drama of the ice age's end, they co-produce the planetary surface with humans. As rhetorically distant this language of rivers is from the language of water engineering, the hydrological and emotional power of large, intersecting rivers is a phenomenon that all those who approach, and who stay to experience over time, must recognize. Still, unlike the infrastructure-dense confluence of the rivers meeting at The Forks, no bridges or dams have attempted to conquer the flows of the Deh Cho (Mackenzie).

In the next and last section of this chapter, I again visit the interstices signalled by the blank spaces in the double artifact. Now the focus zooms into an erratic setting in the central space that is the source of all four outbursts, the once-upon-a-time Lake Agassiz – Portage Avenue, downtown Winnipeg.

V. ACADEMIA'S ERRATIC SYMBOLS

Conatus names a power present in *every* body ... Even a falling stone, writes Spinoza, "is endeavoring, as far as in it lies, to continue in its motion."
Jane Bennett (2010, 2) in *Vibrant Matter*,
citing Spinoza in *The Letters*, epistle 58.

In Spinoza's philosophy, seen through Bennett's keen eyes, erratic boulders have vitality that comes from their own power to persevere, to persist, to strive (*conatus*).[39] While as glacial debris they may simply count as passive data points, as notable erratics, they have "thing-power" or conatus. Their thing-power may affect us yet still exceed the uses and meanings we ascribe (2–3). And so, I find, as they move from glacial drop-off point to cultural stage, the mighty erratics easily carry their geological cachet.

As an ethnographer who roams (and errs) with analytic vigour for a living, I gravitate toward the erratic aspect of knowledge production.[40] When I come across this erratic piece of vibrant matter, I can't help but imagine how ...

A. Erratic Arrives in Winnipeg from
(Another Part of) Glacial Lake Agassiz, 1970s

Men working in a gravel pit on the shore of what was once Glacial Lake Agassiz dislodge a boulder too big to crush. As is their custom, they roll it downhill where it rests. One day, officials

2.2 Erratic boulder on Portage Avenue shares Glacial Lake Agassiz's bed with the University of Winnipeg.

from the University of Winnipeg arrive to the pit and select an erratic for a monument. They hire men and machines to transport it to the university's main entrance on Portage Avenue.[41] A commemorative bronze plaque hints at the intention behind the presence of the fourteen-ton (28,000-lb.) stone that did not arrive entirely by accident. Under the dates 1871–1971, the plaque's engraved words convey the erratic's unusual history: "This granite boulder brought from Glacial Lake Agassiz."

How extraordinary. I am attracted to this queer collaboration between a boulder hitching a ride on a glacier and then when the glacier melts, getting stranded in what will become the province of Manitoba, where it hangs out for at least 10,000 years until some men with machines come along to stand it up on the sidewalk within view of the Hudson's Bay Department Store, a tired icon of colonial power transformed into a material promise of reconciliation, and Manitoba Hydro-(electricity), a postcolonial powerhouse (see figure 2.2).[42]

Clearly, humans speak to each other in the language of erratics. Like glaciers, geo-cultures create meaningful collaborations by moving erratics across distances and differences.

B. Erratic Arrives in Harvard from the Swiss Alps, 1870s

An erratic from the Hôtel de Neuchâtelois, the famous field station where ice age theory made the leap into nineteenth-century European science, makes a transatlantic crossing to Boston in the 1870s. This is the story:[43]

> Upon the death of his father, Harvard Professor and namesake of Canada's most geologically significant glacial lake, Agassiz's son asks his Swiss cousin to journey back to the Hôtel de Neuchâtel site in the Lower Aar Glacier. According to the Mount Auburn Cemetery, the cousin selects a 2,500-pound (one and a quarter ton) erratic, arranges for a group of men to carry it twenty-five miles in their arms to a village where they put it on a wagon to the nearest train station. Officially, it is the Republic of Switzerland that sends it across the Atlantic Ocean where, notably, the erratic's surface, except for the inscription of names and dates 1807–1873, remains untouched by stone-cutter's hands. Its appropriated nature continues to reign intact. [my paraphrase]

And so, to bring this chapter "ice/water on the move" to an end, I collect two erratic stories and put them within the same frame. They move first with a glacier, and after resting in postglacial warmth for thousands of years, the erratics find themselves moving again. In the arms of men hired by university officials and a famous scientist's relative, they are rolled into their new resting places, set in a university cemetery in Boston and at the entrance to my fieldwork base, the University of Winnipeg. And so, glaciers and humans enroll the nomadic magnitude of stone in ongoing projects that entail moving across spacetime by force of direction and attraction if not always with purpose. In this manner, erratics condense meaning through accumulation of ties with glaciers, persons, institutions, and nations. By assuming central roles in these geologic dramas, erratics help humans to celebrate the production of science, aesthetic form, identity, even as they slip into the background, impressive but ultimately mundane props on the urban stages of human-to-human dramas.

And yet, might the romance with erratics exceed any one composition of particulars? They go beyond their intended use by reputation boosters, who I presume use erratics to convey the size, weightiness, and beauty of officially recognized accomplishment. Looked at another way, might erratics recast rather than simply enable humanity's tendency to exhibit mastery over nature? What might an erratic perspective do to this geo-cultural story? Here, I devise an erratic act of ventriloquism.

If I could write like an erratic, I would say this to the humanimals:

> *Our journey as singular entities begin when we are torn from rock formations much greater yet similar to ourselves, stuck in the cold embrace of slowly flowing frozen rivers. We roll and bang across spacetimes, colliding with all kinds of things, beings, unexpected elements: experiencing alterations in mineral composition, weight, shape, colour, texture and surrounded by unknown rhythms and weathers. People, coming close, feeling our patterned surfaces and sizes: some recognizing our persistent character, distinctive liveliness, our vulnerabilities to disconnection and attraction to new, provocative configurations. We find affinities with other glacial actors on geo-cultural stages. Kidnapped, not once but twice, we are set to roam by forces beyond our control and have no desire or means to ever return to our origins. Our integrity remains intact.*

So to speak, erratics are inorganic, nonhuman characters that enact their state of being in spacetime frames dramatically differently from our own. Thinking again with Massey's (2005, 130–8) "Migrant Rocks," erratics provide clues to unravelling the particular "throwntogetherness" formed by glacial action. Their being "here" on Portage Avenue in front of the University of Winnipeg (or beside Agassiz's Boston grave) is not the same as humanity's being "here." This compelling ontological difference energizes the cultural travels of erratics. Appearing as rocks with stories,[44] I'd like to use their living histories to trouble lines that separate animate and inanimate, the lines we tend to think we draw whenever we conceptualize world-making in a scientific frame of mind. The fact is, though, neither humans, glaciers, nor erratics can return to any of the "heres" we had before we burned so much fossil fuel and warmed the atmosphere and oceans. Glaciers that have survived into the Holocene on the poles and in the highest places will succumb to our thoughtlessness should habits proceed apace. What to do? As I move into engineering ice/waters in contemporary spacetimes, I keep this question in mind, approaching geoscience with imagination and as imagination.

3

Systems and Seasons
(1950 and 1997 Floods)

Children in schools learn it as one of four significant moments: precipitation, flow formation on the earth's surface, evaporation, and cloud formation ... The world it seems has cast anchor in the second moment, making it the time of reality when an earth surface can be seen and articulated with some clarity, maps are drawn, properties are demarcated, infrastructure is designed, the past described, the present experienced, and the future envisioned.

Da Cunha (2019, 9) in *The Invention of Rivers*

I. THREE ETHNOGRAPHIC TERMS: TECHNOZONES, BLENDED HABITUS, AND FOCAL POINTS

Most of humankind inhabits one or more cities that have "cast anchor," following Da Cunha, in the moment of river flow, the time of reality when the earth's aquatic nature can be seen, mapped, and engineered most clearly. There might have been other ways to invent this space, as Da Cunha's radical insight in the above quote suggests. But in Manitoban history, the great confluence of the Red and Assiniboine Rivers becomes first a meeting ground for First Nations, then a settler city. By now, sedentary architecture, roads, and rails press the river bends with their needs. When the Red and Assiniboine flow into the tilted prairie, they meet a flood control system designed to hold them to their singular, channel-based forms. This chapter focuses on this encounter in the space of intersecting river basins that stretches outward from the confluence (called The Forks). Three ethnographic concepts organize this encounter in Manitoba and, I suggest, can be used to organize encounters in all flood-prone cities with control systems. Together, they constitute the dash in geo-culture, i.e., what happens, or might happen, in spacetimes that are both geophysical and cultural, deep and anticipatory. They are:

1. *Technozones* (cf. Barry 2006): Dams, spillways, channels, computers, and satellites, etc., have transformed wet prairielands into technozones – infrastructured and regulated force fields.[1] Technozones assume dominion over spatial expanses by establishing systems of common standards, technical practices, procedures, and operational forms. These distinguish what happens within and beside them from what happens beyond their reach. Technozones are neither comprehensive nor determinative, neither bounded nor mappable. At their best, technozones make frozen urban space habitable for city folk.[2]
2. *Blended habitus:* Shared touchstones and tactics of survival are part of a blended habitus – dispositions, sensibilities, and skills (in their less intentional effects and affects). Blended habitus folds in the personal and professional, practices and structures, improvisations and rules.[3] Even those who are protected by, or officially responsible for, operating the technozone rely on the familiarity of despair-wrought knowhow handed down across generations of flood experience.[4] Barry's concept of technozones leaves open the question of culture; blended habitus enters and fills this opening. Ethnographically, I make the case that technozones are geo-cultural predispositions, templates, and technologies that materialize in tandem with vernacular floodways.
3. *Focal point:* To blend science-based technical traditions and vernacular floodways, a hydrogeological focal point is needed. Measurements gleaned from a vast network of river gauges distributed in tributaries throughout the Red and Assiniboine basins must be distilled into official communiques, and official communiques must in turn be reinterpreted so as to make them useful in Winnipeg's variously situated homes and businesses. That focal point is located just north of the confluence. Every flood-prone city needs at least one focal point for effective emergency communication.

This chapter weaves the concepts of technozone, blended habitus, and focal point into two more parts that together provide readers with an overview of the system: II. Modern System Basics (the infrastructural assemblage and its routine operations and rules); III. The Extreme Flood Event of 1997 (improvisational earthworks and the first big test of the Floodway). Rooted in the twentieth century, between the Great Flood of 1950 and the opening of the modern control system, this chapter conveys a confident engineering sensibility. Although not without tensions, the basic arrangement of river-city expectations still seems relatively secure. The moment precedes the disorientation of the climate crises manifested in the extreme flood events in 2011 (see chapter 4) and in 2014 (see chapter 6).

II. MODERN SYSTEM BASICS

A. Two Axes of Flood Control Design

The Ministry of Infrastructure and Transport (MIT) is the provincial agency responsible for flood control, including the operation of infrastructures and surveillance networks.[5] The collection of walls, gates, and channels is designed and operated to block the ice/water flows crossing into the city through the Red and Assiniboine River basins. The giant earthwork-machines are clever enough to dance with the massive force of the Red and the smaller but roguish Assiniboine. (See figure 3.1.)[6]

> *In the south-to-north axis:* At the southern tip of Winnipeg, the monumental Red River Floodway control structure blocks a portion of the floodwaters before they enter the city, shunting the excess into a forty-seven-km artificial grass-covered earthen channel before dumping it back in the river upstream below the Lockport dam.[7] The widely recognized feat of engineering was built in the wake of the Great Flood of 1950 and opened in 1968. It goes a long way to solving the city's problem with the Red, successfully containing the flood of 1997 inside the city (though barely). Central neighbourhoods are also protected by a system of primary dikes that raise the Red's channel walls. There is neither dam nor reservoir on the south-north axis, and the Red River meets no control structure on the US/Canada border.

> *In the west-to-east axis,* the Assiniboine River flows down from its headwaters in eastern Saskatchewan, then across the prairielands toward Winnipeg. Along the way, it meets the Portage Diversion

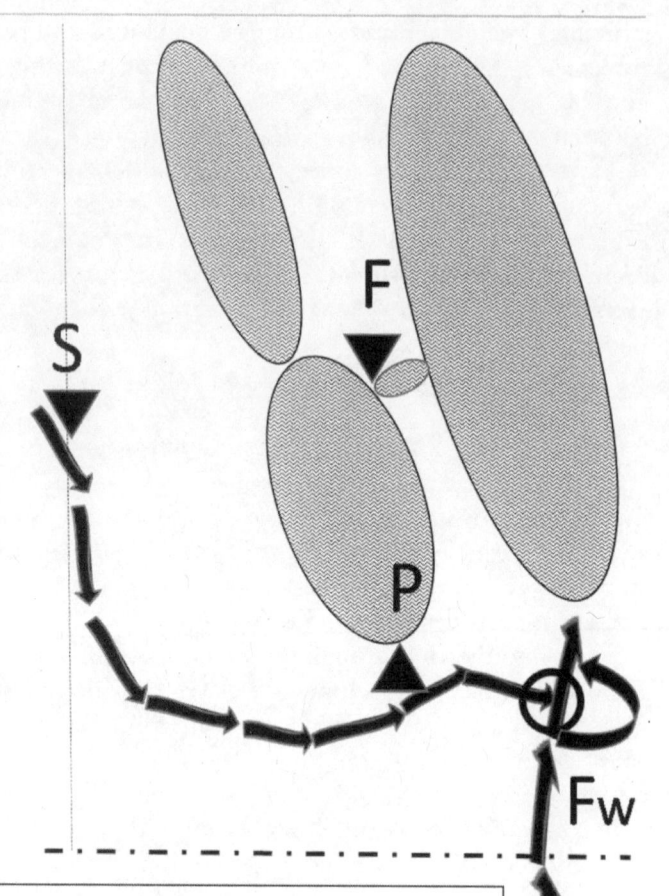

Key to Flood Control System
S=Shellmouth Dam, **P**=Portage Diversion,
F = Fairford Dam, **Fw**=Floodway
Lakes =

3.1 The Provincial Flood Control System relies on four major infrastructures and three lakes. From left to right the lakes are: Manitoba, St Martin, Winnipeg. Lake Manitoba is the largest lake receiving and storing excess water from the Assiniboine River through the Portage Diversion. The name Manitoba most likely comes from the roaring sound of pebbles in the strait connecting the two basins: "the strait of the spirit or manitobau" – from the Cree (Thomas and Paynter 2010, 50) and/or Ojibway languages (https://www.gov.mb.ca/chc/origin_name_manitoba.html).

(see figure 1.2), the most political of four flood control infrastructures central to MIT's mission: 1) the Shellmouth dam and reservoir on the border with Saskatchewan regulates flows east into Manitoba and downstream into the city. Its dual storage capability offers both flood protection and drought mitigation.[8] In flood events, the Shellmouth dam and reservoir system works in tandem with 2) the Portage Diversion,[9] which prevents excess water from flowing downstream to Winnipeg by sending it into 3) Lake Manitoba, which serves as a storage basin. When Lake Manitoba overflows, excess flows out through 4) a dam on the Fairford River (beset with problems; see chapter 4). Between the Portage Diversion and Winnipeg, dikes provide a patchwork of private and public construction with varying age, materials, and maintenance regimes.[10]

B. Essential Supplementary Earthworks

Mounds and Ring Dikes: At best, the flood control system does nothing to protect the suburbs, small towns, and farms that sit along Highway 75, the main artery connecting Manitoba to North Dakota, which runs down the Red River Valley midline. At worst, the backwash that results when MIT gives the command to raise the Floodway's baffles creates artificial flooding for adjacent neighbourhoods on the unprotected side. There are debates and relevant laws regulating the infrastructural outsiders created by this situation. Original inhabitants have been compensated before. But now, if newcomers build structures in the zone just south of the city, the law requires that they first raise an earthen mound on which to build or that they build a protective ring dike around the structure. Flood insurance is not available. Along Highway 75, all the small towns are surrounded by giant ring dikes. Each has one or more openings, like driveways, for people to get in and out in normal times. Earth and earth-moving equipment await floodwater's approach. In 1997, all the ring dikes were closed and raised to withstand the "Red Sea's rough wind-whipped waves."[11]

Sandbags Can Be Moved Anywhere and Everywhere Needed: Despite the ample prowess of the Floodway (actually, a very long, passive, and visually unimpressive ditch), I would argue that the dominant visual symbol of flood-fighting in Winnipeg is the humble sandbag. Like

Wall Street banks, monumental earthworks are too big to fail. They engage the scale of mighty rivers, but their lack of flexibility limits effectiveness. The flood bowl is a vast, complex system. As sophisticated as the surveillance and interpretation of flows are, when the flood enters the prairie surrounding the city and the city itself, its pathways and myriad smaller-scale interactions with winds and roadways cannot be predicted except in the last moment (see below). Which tributaries will unexpectedly bridge a gap in defences? Which dikes will reveal their internal weaknesses in the moment when they are most pressured? As the pattern of flow and failure reveals itself, sandbags, readied for the purpose, can be filled, moved, and heaved onto crucial spaces of hazard by dedicated squads of inhabitants and soldiers. Sandbag maneuvering allows inhabitants to match the local and familiar scale of river action, filling the nooks and crannies of dangerous water infiltration. Moveable sacks of earth elements stack into rising walls shaped to meet the rivers' trickeries. Sandbags are the tools of inhabitant and soldierly action, the folkways of flood-fighting come to the fore.

Winnipeggers are not simply victims waiting for the flood to recede so that they can clean up the filthy mud left behind. They are much photographed actors in a dramatic moving sociotopography of determination that mobilizes relief efforts and fills the archives of memory (Bumsted 1997, 54–5; Thomson 1997; Currie and *Winnipeg Free Press* staff 1997). Sandbags, simple technological bases for action in visual imagery that roams with the rivers, lend expressive shape-shifting power that can shift the historical grounds of blended habitus. In contrast to monumental earthworks fixed in space, impenetrable to ordinary folk, sandbags move as the rivers move while the images of inhabitants who wield them travel through and beyond the storied spaces of disaster.

C. Routine Operations in the Order of Seasons

Monumental concrete and steel engineering structures are certainly essential to flood control. But they can only help the city if engineers who operate them have access to forecasts based on real-time flow conditions. While hard-edged concrete and steel structures themselves may register provincial power on the landscape most emphatically, flood control operations are almost completely dependent on communication infrastructures that link networked points on the flood bowl surface through satellites to computerized sensor systems on the ground. The human–river relationship is a 3D cyber affair. In it, multi-directional, multi-scalar information flows mimic seasonal and sociotopographic variations. In this 3D world of information flows, the satellite-transmitters circling in the atmosphere gather and reflect points across the hydrogeological surface. Calculations use difference from sea level as a quantifiable reference point for the river-position of each individual gauge, thereby converting the different elevations in topography (each of which affects the river flow rate) into seamless data streams. Thus, the bumpy surface of the earth is made flat enough to squeeze into predictive models that guide the operators of earthworks.

I base this basic system overview on a synthesis of a core set of semi-structured interviews and site visits with MIT engineers, forecasters, geographers, and policy staff, past and present.[12] The chapter also draws on retrospective archival accounts of the flood of 1997, the last major traditional flood. I describe 1997 as traditional insofar as it fits fairly comfortably into post-1950 expectations regarding the timing and directionality of flooding, the sense of security that the modern flood control system had been cultivating among Winnipeggers, and the relative confidence in civil engineering before the disruptive questions accompanying climate change take hold of global geo-cultures.[13] After 1997, I suggest, habitus turns. And that habitus *can* turn is due to ongoing, pragmatic, and informed negotiations with the flood bowl's fundamental throwntogetherness (Massey 2005, 140–2) on the part of forecasters, engineers, and ordinary folk.

Seasons mediate river-city states-of-being.[14] Motivated by riverine impulses, expert work moves through seasonal phases that include backstage routines and frontstage action (see below). All the while, engineers attend closely to rivers' streaming signals, the digitized qualities arising from the provincial network of mechanical sensors. In Manitoba, Water Survey of Canada (WSC) data combine with United States Geological Survey (USGS) data on the cross-border Red River and the Souris, a tributary of the Assiniboine. Once a dangerous river crest comes and goes, messes are cleaned up, reparations made, decisions justified, inhabitants act to reestablish the everyday. Stories are born out of dramas that test the reliability of flood habitus blended of professional and folkway knowledge and tied to the viability of concrete infrastructure, virtual data processing, legal regulation, and riverine impulses.[15] All this, roughly speaking, comes together to shape expectation and experience in crisis as well as to shape geocultural memory in the *longue durée*.

1. *Interpretively Sifting through Snowmelt Forecasts:* Snowmelt drives urban floods.[16] Back in the ice age, snow mounds took millennia to compress into ice and millennia for snow/ice formations to melt and break up. In twenty-first-century Winnipeg, a compressed version of this geophysical transformation happens every year. Governed by temperature, wind, and gravity, snowmelt is a geophysical process and an urban event. Around about snowmelt comes breakup, when ice gives way to ice/water. The sounds and furies of a river at spring breakup can become an event that pierces Winnipeggers' everyday lives and marks turning points in the most intimate relationships, as does the "Ass River" (Assiniboine) in Miriam Toews's (2022, 26) personal history published in the *New Yorker*:

> I leaned closer to the window for a better look, my forehead was almost resting against the pane, and it was then that I heard a thunderous sound, a type of explosion, coming from the river path, and I realized that the path was cracking up, that it was the time of the spring breakup, as we called it, and that giant walls of ice, sheets of ice, sometimes more than a hundred feet long and two or three feet thick, were cracking apart and hurtling along on the powerful current of the Ass River, toward the bridges, toward who knows where – the sea, I guess, ultimately.

Once elemental forces of snow and ice/water are in play, expert flood routines switch from watch mode to action/emergency mode. Until then, observational routines fill the calendar of operations. Keen to be prepared, farmers, small-town mayors (some called reeves), entrepreneurs, and service organizations also watch precipitation levels throughout fall and winter. High snow piles, soggy fields, and the depths of ice hardening creeks and rivers provide empirical bases for at-home flood prediction. As snowmelt nears, watchful inhabitants can be fairly certain about whether conditions are conducive to an extreme flood event. For those who know how to read the language of ice/water and snow, spring floods come as no surprise. They watch thermometers. They scan the skies for storm clouds that can throw off predictions by adding too much rain too quickly on top of the melting snow. The parameters of expert and folk surveillance attend to these basic geophysical dynamics: The higher the snow, the faster temperatures rise, the later in spring temperatures rise, the greater the probability that the Red and/or the Assiniboine Rivers will flow over (or under) their embankments and spread like a sea (or lake) over the flat farms and cities of the prairie. All in all, however, whenever extreme flood events occur, the effects and outcomes depend on where one is in the flood bowl.

Most people inside the city's protected zone may have the luxury of ignoring such signals. But those on farms, near eroding embankments, or in vulnerable communities and industries pay close attention. If you live in a floodplain that is outside the central zone of protection, you are vulnerable and have to watch and listen for what's coming.[17] Grains have to be moved out of the valley early in December, a process that takes two months. Utilities, Manitoba Hydro, the railways need lots of time to prepare effective action. By February and March, farmers have to figure out where they can move their animals. Later, they might decide that a temporary dike might be wise. That takes a couple of weeks to build. Should you spend the money? Inhabitants walk the fields, but as winter sets in they also start paying attention to regional forecasts. The closer the time to snowmelt and the more ominous the threat of flood, the more important a wider set of regional data becomes.

"Interpretively sifting" through reports and images of engulfing waves coming via radio, newspapers, television, and websites, farmers and homeowners seek regional information most relevant to their particular spot in the universe (Derrida and Stiegler 2002, 42). Importantly, the prairie bestows the gift of time. If and when the Red and Assiniboine Rivers unleash their powers, they do so from a distance. Ice/water flows build up after breakup, but even those that gather speed and turbulence in western mountain slopes still have to cross the flat distance to reach Winnipeg. Slow, but inexorable, the prairie gives people time to fill sandbags, pile them up on embankments, and in the end, if needed, evacuate persons and animals.

The seasonal rhythms of ice/water and snow produce emotional suspense that governs the work lives of snow-watchers and provides a perennial rationale for the continued support of provincial experts responsible for flood control at MIT. Ministry experts may live in a semi-autonomous, technologically enhanced universe of river engineering structures, surveillance, modelling, decision-making, and formal public communiques, but they too must watch the ice/water and snow according to season. They too interpretively sift through the empirical signs of aquatic excess gathering across their terrain of responsibility. Like ordinary folk, they may also interpretively sift through official news reports and home-base-field-observations to see if and when the flood will personally affect them (see chapter 4 for the story of Ron Richardson and the zigzag dike).

But experts spend most workdays removed from field observation. Tucked away in often windowless offices and meeting rooms papered with maps inside government buildings, they watch computer screens across which seasonal cycles are converted and modelled into digital data streams flowing in the "real time" of laboratory clocks. The experts determine which data to count as facts, which facts count as news, and which news should trigger calls for citizen and military emergency assistance.

The tempo of suspenseful action is elastic, despite the continuity of data feeds into MIT. Data collection, interpretation, and decision-making slow down and speed up in seasonal rhythms. Most of the year, surveillance involves routine data gathering and waiting. Only when situations become more urgent do minutes and hours count. In emergencies, hydrological knowledge (how much water is flowing into the bottom of the bowl and from where) and decisions about hydraulic engineering (when and where to reroute excess flows) become entangled with institutional mechanisms for disaster response that entail a different set of routines.

MIT's surveillance routines for spring flood prediction start in the quiet of fall, when river and reservoir levels are purposefully kept low. New technologies enhance technical traditions of engineers, offering new ways to manage suspense leading up to snowmelt. For example, as soon as the ground freezes, MIT does airborne gamma radiation surveys. Adapted for civilian use, this military technology composes the "spectral signature" of surface terrain features.[18] Gamma radiation technology takes advantage of the geophysical fact that moisture decreases the reflectance capabilities of soil. The more ice/water in the soil, the lower the radiation readings. It thus provides a geovisual reference that can be compared to averages that have been calculated for particular locales. Gamma radiation measurements thus stretch the sense perceptions a person may get from walking through a soggy field. Once frozen, a field will not reveal how much ice it holds to someone walking upon it. But with gamma radiation data, it is possible to penetrate and geovisualize the top layer of frozen ground. If people learn that there is relatively little moisture stored in and on the ground going into spring, while there is no call for complacency, they might be considerably less worried about an extreme flood event. For among the grains of sand, silt, and rock under their feet, there would still be room to hold floodwaters. And thus, technological revelations can lighten or intensify suspense. And yet, as Jody Berland (2009, 223) points out in her analysis of the first TV reporting of weather satellite data in the early 1970s: "What we can see and where we see it are not necessarily commensurate with what we can predict. Farmers and drivers still suffer incalculable damage from the weather's wrath. Eluding the optical and digital extensions of human vision, weather still outwits the computers." Into the gap between visual data made public and the sphere of unintended river agencies walk the flood forecasters.

2. *With Increasing Precision but Never without Scientific Uncertainty, from Outlooks to Operational Forecasts:* During the cold slow winters, MIT staff produces a series of Flood Outlooks. They are best guesses based on various data sets that have been entered into predictive models. The models calibrate current events with past floods.[19] When winter first sets in, MIT does a second gamma survey of soil moisture and adds it to the fall data. As winter proceeds, they quantify cumulative snow pack using radar. They add readings from provincial gauge networks that measure precipitation and temperature and river flows and levels. Thus, data from these and various other technologies are layered together to anticipate spring floods across the province.[20]

Flood Outlooks give emergency response managers an impression of what they may be faced with when ice, snow, and rivers start to move. Forecasting runoff is the first problem. What percentage of snow will run off in each sub-basin?[21] If it looks like it's going to be a bad year, MIT constructs a January outlook. They continue to re-evaluate provincial flood probabilities, publishing adjusted outlooks in late February and late March. Speculative yet practical, Flood Outlooks are open to three kinds of uncertainty: those due to weather patterns, to sampling error, and to assumptions built into the model. (Models are black boxes that intervene between sensor-measurements and interpretations.) In the winter months, engineers and forecasters simply cannot know how much future precipitation there will be, how fast the snow will melt, and if, how, and where ice jams will occur and make floods more likely. And so, province-wide assessments of water level variation and risk are inaccessible. And like the geoscience calculations of Lake Agassiz isobases (see chapter 2), questions regarding sampling reliability may add to uncertainty. E.g., how many points are sufficient from direct and remote readings to establish a valid appraisal? In other words, outlooks are based on probabilistic decision-making.

Operational Forecasts, in contrast, are constructed in the thick of events; they primarily deal with concrete data, not models. As spring advances, unfolding interactions of earth, ice/water, wind, and temperature can be read directly in the flows and stages of rivers and their tributaries. The hydrographic network measures and communicates flow rates from particular gauge sites and tracks crests as rivers move through the bowl. The data is interpreted in relation to comparator floods from the past. Operational forecasts are more specific than outlooks, providing the best estimate of the moving flood crest. *As time to flood shortens, the more precise and certain forecasts will be.* This is a principle grounded in habitus. Unlike outlooks, forecasts are no longer based on how much rain will come in the future. The melt is done. The runoff is done. The Moment of Truth has arrived. How are the tributaries reacting? How much runoff are they receiving? You can actually watch digitized versions of sub-basin flow, which, thanks to cross-border agreements, combine USGS and WSC weather data. The stakes of forecast accuracy rise as the time horizon shortens. Once runoff starts, MIT puts out daily sheets. Forecasters can start to route the flood while still worrying about the possibility of rain.

But even though uncertainty related to modelling drops out, sampling errors persist. Errors are produced by variation in topography, equipment, and data-gathering conditions. Gauges have to be placed and protected from differences in sun, shade, or wind. Weather can still add uncertainty. Geophysical factors related to the interaction of temperature, ice/water, and snow further complicate flood risk calculations. For example, there is not necessarily a simple relationship between temperature and snowmelt. If spring rains fall on snowpack, it can increase the rate of snow melt and hence runoff. In the case of a cold spring, however, the snowpack can dissipate by sublimation – snow crystals vaporize, bypassing the water phase, and dissipate into the atmosphere, reducing flood risk. And too, if snow doesn't cover the ground sufficiently, rain percolates into the soil, freezing more deeply and blocking spring meltwater from flowing through. This increases surface runoff into rivers.

Even if the forecasting team has a grasp of river flow rates and interactive complications, when data analysis steps off solid ground into the arena of extrapolation, good predictions may benefit most from forecasters who command local and historical hydrogeological knowledge. Blended habitus cultivated in the landscape as well as in ministry offices can become crucial when operational forecasting enters emergency mode. There are generational differences in how field knowledge versus modelling prowess is regarded. At least among the men I spoke to, younger forecasters trained to rely on computer models rather than sensibilities cultivated from a life of streamwalking and those who circulate into Manitoban forecasting offices from international circuits tend to have less confidence in the traditional blend of habitus that summons field experience when in emergency extrapolation mode. Economic pressures also influence the blend of expertise. In the 1990s, significant layoffs at MIT weakened the institutional and local knowledge base in ways that could never be fully recovered. One might say, then, that neoliberal management styles shifted the technical practice and politics of forecasting.

For floodplain inhabitants, the implications of forecasting reliability are deeply personal, affecting individual homes and farms. Sometimes one might think things are going well, but conditions might take a turn for the worse, a forecast has to be revised upward, and the person gets caught unprepared. That's why blended habitus includes the "better safe than sorry" principle: *If you are building a dike, always add to freeboard (that is, make it higher) to account for uncertainties.* In any case, vulnerable inhabitants are rightly ambivalent about forecasting, which, after all, is not an exact science. Forecasters have confidence even when there's some degree of uncertainty. There is always uncertainty. As retired forecaster Alf Warkentin, a man who has walked in many a Manitoban streambed, puts it: "Inhabitants shouldn't read forecasts like the bible but neither are they trash."

Forecasts are weather-dependent up until almost the very last moment. If conditions take too long to resolve themselves, it may be too late to craft a sufficiently reliable forecast. Forecasters may get swamped with data, but they nevertheless have to wait for hydrogeological processes to clarify. This produces some tension in government agencies, which tend to prefer acting more proactively. Unavoidable differences between real-time river action and the lagging interpretations of incoming data streams are a feature of geo-culture that demand a sophisticated yet interested skepticism among experienced inhabitants. And following Berland, the skepticism toward forecasting should not simply be about the contest between science and weather but also about how weather and the "compelling televisual rhetoric of observation and mastery" within which it is conveyed enters "the representational landscape of Canadian culture" through the apparently "neutral instrument of North American surveillance." Welcome to the technozone – the wet prairielands transformed into the regulated, infrastructured, and mass-mediated force fields that make the city habitable (cf. Barry 2006).

3. *Quandaries of Measuring Snow for Operational Forecasts:* In 1997, there had been a blizzard and deep freeze throughout the Red River Valley. In early April, the Red River rose up out of its channel to flood Grand Forks, North Dakota. Icy, sewage-laced water poured into streets and basements, shorting electric wires and igniting a blaze that burned down the city's historic centre. The flood had not yet reached Winnipeg, where Alf Warkentin waits, one of the forecasters responsible for producing updates. On 5 April, he speaks to the *Winnipeg Free Press* about the difficulties of measuring snow in a blizzard. In 2014, I come across *A Red Sea Rising: The Flood of the Century*, a commemorative coffee table book published by the newspaper that includes behind-the-scenes reconstructions of flood action in great personal and technical detail, including Warkentin's (Currie and *Winnipeg Free Press* staff 1997, 5).[22] In it, he shares the kind of technical puzzle that can thwart systematic data collection with the journalist:

> The blizzard presented a special challenge to Warkentin because it dispersed the snow haphazardly. "In a blizzard, calculations can easily be out 50 per cent," Warkentin said. "Environment Canada studies show that." Part of the problem is that many of the precipitation gauges in the province have been automated in recent years, and they aren't as accurate as manual readings. But there are problems even with manual readings. "*Where do you measure the snow after a blizzard?*" Warkentin asked. "*At the top of the drift? At the bottom? Or in between?* And we had at least four significant blizzards last winter."
>
> The accuracy is crucial. Small errors in snow water content can easily throw flood forecasts off by up to 100 per cent. [my italics]

How to design, finance, and place gauges so that data can be reliably collected and interpreted in the midst of a blizzard? The question poses a decision-making problem that is exacerbated by radical throwntogetherness. If systematic, if one were standing in a snowbound place at a given gauge, the fix would have to entail making an arbitrary decision. Where do you measure? Measure here, measure this way. In their analysis of the whole range of post-flood activities, civil engineers Simonovic and Carson (2003, 352) argue that new stations in improved hydrometric and meteorological networks can overcome vulnerabilities by supplying "data on temperature, rainfall, humidity, wind, global radiation, soil moisture and soil temperature." Sidelining the gauge debates, they nevertheless add a caveat: "Reducing the number of gauging stations in the Red River Valley is not acceptable." And too, they target Canada's conservative "data culture" – by which they mean the unwillingness of agencies to pay the cost of sharing – which in 1997 left users in the Red River Valley with fragmented and incomplete information.[23] And thus, from the vagaries of wind-swept snowbanks, through a surveillance system always in development, to the vagaries of data dissemination, techno-cultural vulnerabilities are bound to introduce error even into "concrete data" informing Operational Forecasts.

In raising measurement issues underlying Operational Forecasts, I do not intend to undermine confidence in their usefulness during approaching flood events. It's simply a basic geoscience field problem. The situation is not unlike the 1960s confrontation between Kupsch, who argued that geoscientists were relying on insufficient numbers of data points needed to establish isobases, and Elson's counterargument that it is possible to be pragmatic about the definition of sufficiency (see chapter 2). In the balance between comprehensiveness and extrapolation, history takes Elson's side. Whether the methodological question is how many data points are needed to establish a scientific fact, as in that earlier exchange, or where and how data points should be collected in a shape-shifting topography, the questions simply need to be answered by the best abilities of those involved. Habitus builds confidence in systems through principles such as: *As time to flood shortens, the more precise and certain forecasts will be*. Habitus also accommodates measurement problems in principles such as: *If you are building a dike, always add to freeboard to account for uncertainties*. Forecasters pin the arc of less to more precision by switching genres from Flood Outlooks to Operational Forecasts. They also translate standard deviations from the language of statistics into a locally familiar form of material culture by using the term "freeboard." The technical language of flood control works best, it seems, when adapted to local (and spacialized) turns of phrase.

III. "NUMBER-FEET JAMES": CONFLUENCE AS FOCAL POINT OF COMMUNICATION IN EXTREME FLOODS

When fearful flood emotions swirl synchronously, many old-time Winnipeggers who live along the Red or Assiniboine Rivers listen for news about one particular datum on the radio – a historic river spot immediately north of The Forks: "Number-Feet James," a measure and symbol of the city's geo-culture. An index of the force of the combined rivers, Number-Feet James provides inhabitants with a focal point for coordinating and interpreting water flow across city and provincial spaces.

A. Risk Interpretation:
What Everyone around Here Knows about Moving River Crests

I kept hearing announcements about "Number-Feet James" on the radio. It was a cipher, a sympathetic link communicating something about flood threats. Repeated enough times in French and English air waves, I pursued its meaning. It's just that kind of thing ethnographers go after – something everyone around here knows but you. A measure, a meeting point, Number-Feet James is a code that outsiders like me have to crack in order to participate effectively in flood discourse.[24]

Number-Feet James is a continuous reading of flow on the Red River in the first bend after it merges with the Assiniboine, by the James Avenue Pumping Station, just downstream from Alexander Docks.[25] Fixed in one spot, the flow reading unifies and orients inhabitants concerned about a present or possible flood. The spot reflects the sum of converging rivers. To make use of the code in practice, interpretation is required. Inhabitants need to master geo-cultural knowledge of the overlapping river basins. For one, inhabitants need to think relationally in order to decode the index. To make it relevant to their own situation, they compare and contrast their current situation to the current situation at James. Inhabitants must understand how to determine the difference between the combined rivers at Number-Feet James and the level in the river reach nearest to them. Figuring out the volume and force of the water most relevant to each situation involves taking into consideration the fact that rivers (much more than lakes) slope along a gradient. In times of flood, forecasters closely track and publicly announce the moving crests of each river. Flood crests may move independently or synchronously; the flood may come on one river and not the other. To be useful, the Number-Feet James has to be interpreted accordingly.

Second, to interpret Number-Feet James, inhabitants think of their personal flood histories relationally. Memory of key details links past floods to the current situation to provide the basis for interpreting how much water and how fast water is coming. Knowledge of past floods may come from personal experience or have been handed down across generations. For example, a family who lives on the banks of the Red knows, say, that in the last flood the river reached three steps up to the front door and that was equal to x Number-Feet James. Thus, Number-Feet James is a communique issuing repeatedly from one site that can be combined with knowledge gleaned from past experience of particular, familiar landscapes to enable ongoing risk assessment. The longer the living intergenerational memory a family has available for their site, the more precisely they can apply the general information they receive from forecasters.

The world has changed around Number-Feet James – the waterfront architecture, measurement and communication technologies, law and policy – all different but for this singular spatial index. This marked persistence in the cultural orientation toward flood risk shows that a technical measure can be uniquely local in language as well as in location. Although it would not be feasible to export "Number-Feet James," one might take the idea to other river cities and search for named locations where flood watch traditions like this could be started if they don't already exist.[26] As flooding intensifies with climate change, the integration of shared reference points in the spatial histories of river cities could be an important communication tactic of blended habitus.[27]

B. Finding the Networked River Gauge on James Avenue

Measurements of Number-Feet James layer into the hydrological panorama of twenty-first-century data streams in more than one way. As technological systems advance, new pathways of data delivery tend to join, rather than replace, old pathways. In this sense, hardware collects in landscapes as data collects in the hydrogeology corpus (see chapter 2). From sensors bobbing up and down in the river to satellite-mediated data streams, to data processing models, the central system of hydrological measurement has been digitized. And yet, river level data from the Red at James Avenue is also sent by phone; related technicalities are also handwritten. The redundancy guards against loss and is also a way to build flexibility and convenience into systems that rely on the measure at Number-Feet James. That said, in some circumstances, people who carry out and communicate measurements may unintentionally drift from old technologies to new, using whatever happens to come in handy (Kane, Medina, and Michler 2015). As it dialogues with riverine agency, a panorama of hydrologic data may well travel technozone space more loosely and heterogeneously than one might imagine. Here's how I came to learn about this (a fieldnote-based account):

> Early on in Fieldwork: I find Debi Forlanski in the back offices of the old train station at The Forks.[28] This wry, passionate yet irreverent supervisor of the Manitoba branch of Water Survey Canada gave me my first lesson on flood ("toilet") bowl hydrogeology. After perusing maps and photographs that helped convey the experiential dimension of her survey work, we get in her truck for a quick drive to the gauge at James. There it is, sitting in a weedy patch of bank, perhaps the single most important spot in the "meshwork" of official and folkloric micro-practices that help to establish sovereign territoriality.[29]
>
> Up close, you can read information on the stick-on label with Environment Canada's maple leaf logo. But like all the provincial network gauges, this one is housed in unobtrusive grey metal cabinets. Unlike the water palaces and pumping stations of late nineteenth- to early twentieth-century water infrastructure, which were built with ornate style to impress the citizenry of their importance, today's river surveillance infrastructure relies on an intentional aesthetic of backstage obscurity.[30] The work of public education is carried out elsewhere.
>
> Across the street in a closet at the corner of the James Avenue Pumping Station, another electronic level transmitter (a measuring tape) drops into a culvert under the street and into the river.

Its readings of Number-Feet James are sent through a phone line that dials into the city every fifteen minutes.[31] The phone reading is used to guide operations at St Andrews Locks and Dam downstream, which maintains a fairly constant river level during Winnipeg summers for aesthetic and recreational purposes. The St Andrews Dam is also used to lower the river level in October, a routine designed to reduce winter ice formation and increase the channel's capacity to contain melting ice and snow in the spring freshet. (See more on seasonal operational routines and their intended and unintended consequences in chapter 4.)

In contrast to the Floodway upstream, which shunts water around the city only during peak flood events, i.e., in emergencies, the dam downstream modulates seasonal river levels so that city life can proceed the way it should in normal times. But the closeted gauge at the corner of the old pumping station has been reduced to a supplemental one for the city. This kind of redundancy is created by technological advance (the system has effectively been digitized), infrastructural drift (the system has shifted into digitization but not completely), and fealty for the tested and true (the sensible that can be read and written down by humans if needed).

The river gauge network is maintained by Water Survey in a cost-sharing arrangement. Manitoba Hydro pays for almost all the gauges in the north, where most hydroelectricity is produced, and MIT pays most of the cost of gauges for the south- central, where they are responsible for flood control (see Figure AII.1). Back on the weedy riverbank, the satellite-linked gauge in the cabinet sends and integrates the Number-Feet James readings into a vast hydrometric network of gauges set up throughout the flood bowl (see figure 3.2).[32]

A partially buried metal tube encloses a wire carrying readings from pressure sensors that register the height of the river into a control box plugged with various other wires that sends the signals up to a satellite. In its turn, the satellite links down into the offices of MIT and Manitoba Hydro. The cabinet is one of hundreds that connect information bits encoding river flow to office computers via satellites. A hand-written log with notes contrasts observed and corrected data, and any maintenance issues register the visits of human technicians like Forlanski.

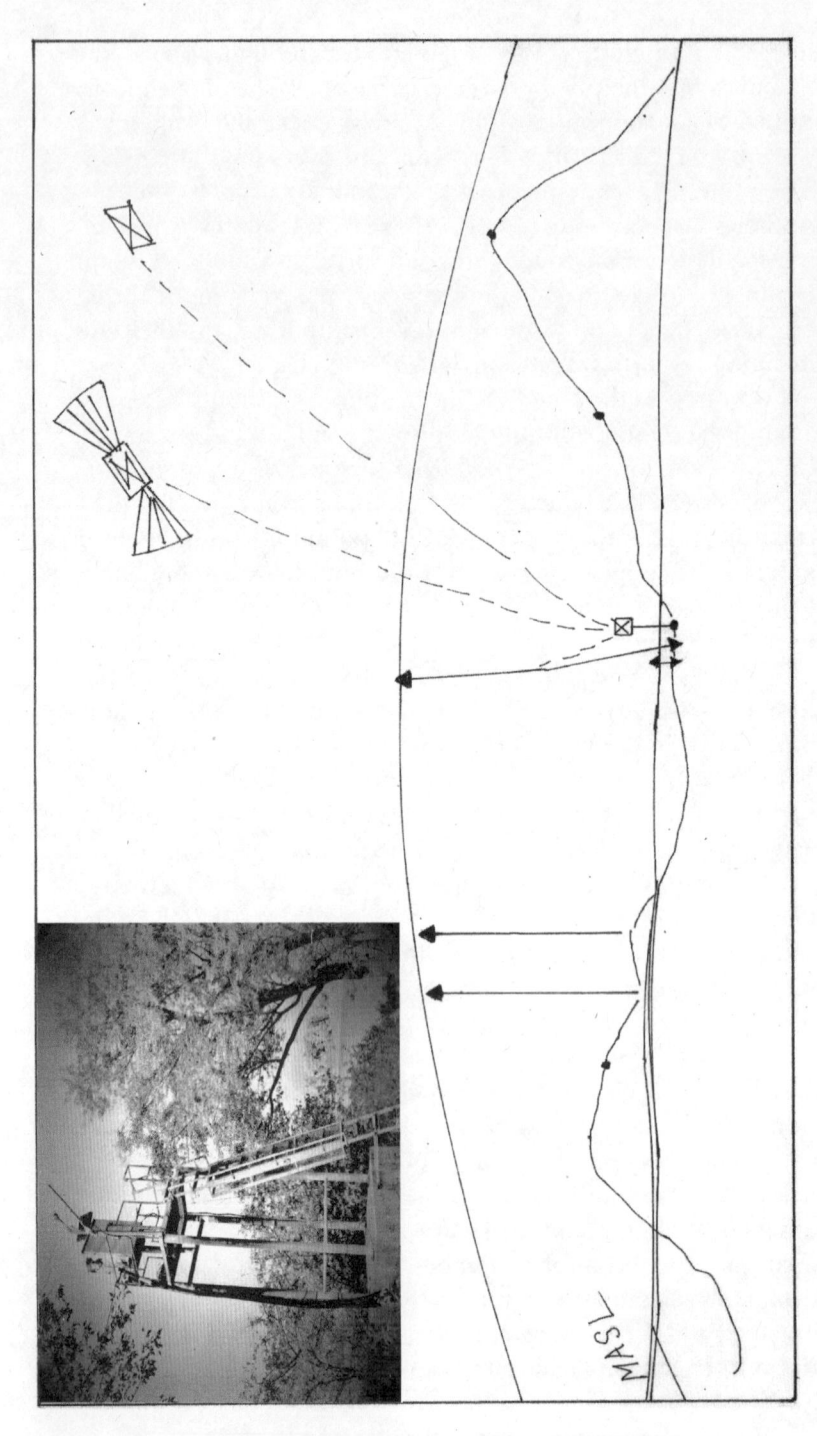

3.2 The hydrometric river network relies on surface-water-to-satellite data flows measuring and computing water levels and metres above sea level (MASL). Photo by author shows gauge near the Red River Floodway Control Structure south of Winnipeg.

In flood emergencies, gauges may get added, cut, mounted incorrectly, or left in disrepair. Like cell towers in earthquakes, river gauges can be pushed down or out of satellite alignment by wind and water, during or between flood events. Gauge network problems can cause some areas to drop out of surveillance range and may distort general forecasts. They need to be constantly checked. Sometimes in the middle of bad floods, Debi Forlanski and her team have to travel out to distant locations in difficult conditions to fix gauges. People beyond the meaningful range of Number-Feet James may depend on river-level measurements at gauges far from the city. In flood emergencies, Forlanski's responsibilities out in the field contribute most directly to public welfare. She tells me that her role in such critical events is what makes her rather mundane bureaucratic job as supervisor deeply rewarding. This ethnographic fact illustrates the meaningful, human dimension of even highly sophisticated technical systems.

Experts may blame gauge problems for bad forecasts or other operational problems in technical reviews. Such problems may trigger contentious public discussion within and between provincial jurisdictions. For example, when Manitoban farmers' land and crops were badly damaged by the 2011 flood, they aimed their wrath at faulty gauges in Saskatchewan, the neighbouring and higher elevation source of the Assiniboine (White 2012). But with all its widely distributed hardware and software, the gauge network has complex vulnerabilities. When it fails despite all the maintenance effort, it may seem like the flood control system is at the mercy of forces as quixotic as the snow and wind.

C. Political Engineering in Language and Law

Linguistic traditions associated with measurement are an important dimension of blended habitus. Place names that index a relatively stable site, like the stretch of river just beyond the confluence of the Red and Assiniboine Rivers, can have special importance in an otherwise fluid landscape. Place names can hold cultural wisdom (Basso 1996) in ways that support the functioning of non-traditional infrastructures. I suggest that Number-Feet James is a form of intergenerational wisdom, unique to geo-culture. It helps inhabitants to navigate hazardous sociotopographies. As stated above, the measure demonstrates a persistence in cultural orientation toward flood risk that is uniquely local in language as well as in location. But there is more to this focal point than its contribution to emergency action. Indeed, the functionality of Number-Feet James is inseparable from the material history of Winnipeg's flood experience. Without it, flood memories that sustain the interpretive prowess of inhabitants as they move into future flooding worlds might unravel. Or at least, they could lose important shared reference points.

There is evidence that the collective orientation provided by Number-Feet James is recognized as a communication tactic by experts as well as the folk. That engineers measure Number-Feet James in imperial feet rather than, or in addition to, metres attests to its significance as a connector of situated histories and technical traditions. Imperial measurement perseveres despite the conversion to metric units in Canadian culture more broadly. Winnipeg engineers switch from one to the other with precision. MIT's engineers work in metric but report Number-Feet James in feet. They publish hydrometric and distance data in both feet and metres. Inhabitants are comfortable with getting their weather in degrees Celsius, but flood interpretation is another matter entirely. Thus, technical terms come to signify local culture through the persistence of global words from a past that sticks, a linguistic assemblage no less distinctive than the backstage techno-cultural assemblages of maple-leafed gauge cabinetry that house riverine data flows.

The iconic measure of Number-Feet James coordinates the knowledge and actions of experts and ordinary inhabitants and is thus a key element of blended flood habitus. Flood control experts such as forecasters, operators, politicians, and journalists benefit from professionally cultivated habitus; their technical traditions become especially salient in emergencies. Although tailored to government and fieldwork sites, expert technical traditions are like folk traditions. They are also encoded and transmitted through narrative performances of recalled perception, knowledge, and action (Scarry 2012; Kane 2016).

1. *The Role of Number-Feet James in Operational Rules:* The well-known #1 Rule of Operation of Red River Floodway Control Structure guides emergency MIT decision-making before, during, and after floods. Proper enactment of the rule requires constant monitoring and calculation of water levels to assure that the Number-Feet James is below 24.5 Feet James, the technically fixed "natural level."[33] But although folk and expert language and practice may both orient to

the same indicator, the way they use them diverges. For example, even if the #1 Rule uses Number-Feet James in a way that ordinary local people understand, the concept of "natural level" refers here to a complex calculation: the height of the river that would have occurred in the absence of flood control works, with the level of urban development in place at the time of the construction of these works. Wow. This guideline for action in the present is based on a definition that invokes an environment that has not existed for centuries. The calculation upon which it is based may well enter into public discourse when there is conflict over operational decisions, but the calculative process by which the measure of natural level is arrived at must be elusive at best for most people, including this ethnographer (see analysis, chapter 5).

Suspense grows as water rises and engineers consider abandoning Rule #1 for #2 (in a major flood) or #3 (in an extreme flood). Negative political repercussions decision-makers may face escalate in tandem with stepped-up risks. In any given moment, engineers might have to decide, for example: Is it best to operate according to rule #2 and cause river levels to reach higher than 24.5 feet, the legally defined natural level? Or rather, should they stick with rule #1 and not allow the Red to go higher than 24.5 feet? The more engineers operate the system so as to protect people inside the protected areas within the technozone (i.e., keep the level of the Red at or below 24.5 feet no matter the oncoming flood), the more they put the people outside the protected technozone, the infrastructural outsiders, at risk. And so, with this question – whether they should follow Rules #1, #2, or #3 – an engineering decision formally becomes political. Decision-making moves up the government hierarchy from technical to political actors, from the Ministry of Infrastructure and Transport (MIT) to the office of the first minister of Manitoba, the premier.

The rules encode bureaucratic pivot points. The framework for decision-making is no longer simply about hydraulics but rather about how to distribute the emotional affects and negative material effects of human-river alliances fairly or efficiently. Whatever the outcome, one thing is sure – in the coordination of infrastructural maneuvers, Number-Feet James persists as an organizational focal point. At least it persists while MIT is following Rule #1. One fact that is apparently not well understood by the populace: Number-Feet James tends to drop out of the official texts in guidelines #2 and #3.[34]

That operational decision-making rules and processes link the urban focal point, Number-Feet James, to cyber-earthwork systems timed to a traditional, science-based sense of season is no doubt a generally effective, tried and true way to run the technozone. The logistical power generated at the provincial level can feed back to reduce resulting harm to infrastructural outsiders (see chapters 5 and 7) and give some support to municipalities that are left to solve the mundane geotechnical problems in the neighbourhood, meso scale (see chapter 4). There's a lot of patching up to do to make things right after a flood emergency and, too, as an unintended consequence of routines designed to prevent them. For some towns, like those on Highway 75 south of Winnipeg, Number-Feet James is irrelevant. For example, floods reaching Winnipeg will have already engulfed the town of Morris. This is why its mayor walked around with a card in his wallet detailing the height and time-length of Morris flood levels in 1979, 1997, 2006, 2009, 2010, and 2011, both with and without dikes to protect it. The towns outside the technozone thus may create their own focal points and frames of reference to help assess oncoming risk.

The logistical powers of water channelled through the technozone impose ongoing contradictory consequences on provincial territory – but imposition only works up to a point. Systems and seasons are always changing and only partially coordinated. In the next chapter, I show how routines of flood control engage the contradictions of seasons. I argue that sustainable logistical power comes not from domination and the erasure of contradictions but rather from successful navigation through them.[35] The characteristic super-dynamic tension between order and disorder in urbanized, intersecting floodplains provides evidence for this argument, especially when rivers rush beyond the channels meant to contain them or when channels collapse because the system withholds the water needed to keep them standing.

4

Falling Out of Synch (1997 Flood)

1. ROUTINE AND EMERGENCY MISMATCHES

Everywhere in the world, scalar mismatches inevitably disrupt the smooth functioning of engineering systems. Even in the relatively stable heart of Manitoba's flood control technozone, seasonal operations produce uneven effects that are geophysically uneven and politically obscure. This chapter shows how blended habitus operates through experts and ordinary folk both routinely and in emergency mode to overcome, ignore, and/or perpetuate resulting negative impacts.

Routine mismatch: Flood control is designed and operated by the province at the spatial scale most relevant to the protected areas of the technozone writ large. At the neighbourhood scale, provincial operations can produce contradictory, yet predictable, effects on the structural integrity of the myriad networked interfaces between water and earth that organize urban space. When riverbanks routinely fail, especially in the bends of meandering rivers, city engineers work to patch them up where they can and watch the banks slowly succumb to the rivers where they cannot. This rather mundane engineering (in)action is a little-noticed but basic aspect of provincial territorializing. If, as Elden (2010) argues, territory itself is a political technology,[1] the routine engineering projects that patch systemic mismatches are an implicit politics, immanent in future incidents whose root causes are too complex and entangled in the way things are to resolve. But might mismatches such as riverbank failures – internal to and contemporary with the protected centre of the control system – offer tactical sites for discovering, following Byrd (2011, xxx), a "diagnostic way of reading and interpreting colonial logics"?

Emergency mode mismatch: Major provincial control structures like the Floodway can also succumb to the perils of scalar mismatch. These perils tend to show up in emergencies. As the force of floods intensify with climate change, mismatches between assumptions made in modernity's mid-twentieth-century designs are increasingly falling out of synch with end-twentieth- to twenty-first-century flood extremes. Such is the case in the midst of the 1997 Flood of the Century, when the oncoming Red River looks like it is going to exceed the 150-year PMF (probable maximum flood) built into the structure's protective wings. An amazing feat of collective action to ward off disaster by the construction of the "Z-Dike" assumes a storied place in public memory. The Z-dike, a high-profile negotiation between city inhabitants and Red Sea, stands out from the prairie farmland around it as a territorial artifact and political technology.

II. PROPERTY REGIMES AND THE CONTRADICTORY EFFECTS OF SYSTEM OPERATIONS[2]

Well within the interstices of the flood control technozone, one river reach shows how macro-scale provincial priorities of the flood control system can have negative consequences at the meso-scale, i.e., in the interlinked landscapes that compose the protected heart of the city.[3] Indeed, although varying in kind and effect, geophysical problems caused by provincial flood control may be commonplace, taken-for-granted features of macro-scale infrastructural systems more generally. Neighbourhood scale riverbank problems generally do not migrate to the foreground of urban politics and culture. In this sense, they are mundane, apolitical, or neglected aspects of material life. As precedents for territorial disasters build in the meso-scale of interlinked riverine neighbourhoods, MIT may or may not be able to enact structural repair. Between the first and second Red River bends south of The Forks, neighbourhoods on opposite banks provide contrasting cases. I return to my fieldwork neighbourhood, the side of the river where a successful repair fortified the curved green wedge of flood-prone land at the tip of the bend, the public park along Lyndale Drive (described in chapter 1).

A. Norwood Neighbourhood, Red River's East Bank

Even within the cities protected zone, floodwaters can and do dramatically change the landscape, impressing inhabitants with water's power. Indeed, pieces of the riverbank adjacent to Lyndale Drive are prone to collapse. On my daily walks, the push of the wind-ruffled silvery grey water moving through the channel is a source of pleasure, puzzlement, and study. Looking carefully at the pattern of rock, earth, plant life, timber, cement, metal, and asphalt reveals that there has been much tending of the water-land interface (see figure 1.3). As visually unobtrusive as the material culture of the riverbank is, if one looks for mute evidence of the mundane micro-actions that reinforce today's sedentary settlements, there is much that meets the eye.

One day, I am happy to see a group of people planting trees in a cleared slope along the reach of the Red.[4] It turns out that they are employees of the city government's Naturalist Service Office working on one of the final stages of the Lyndale Restoration Project. They are planting clumps of riparian trees (such as basswood and American elm), bushes that will reseed, and seedlings held down by hydro-mulching for moisture retention and erosion control. As the plants grow, they will help to secure the engineering project that has restructured the bank below ground. If I am to learn more about that effect, the botany squad tells me I will have to talk to the geotechnical engineer in charge of the project, Kendall Thiessen.[5]

1. *Ancient Clay beneath Our Feet*

It is 10 October and already cold out by the river. When I meet Thiessen at the site, I learn that he has recently completed a PhD in geotechnical engineering from the University of Manitoba. The Lyndale Restoration Project is his dissertation project. The lower bank has failed historically at this stretch where the river comes out of the bend into the channel wall. In the 1970s, the city government built a wall of logs to try to hold back the clay, which tends to slide off into the channel. Its engineering workforce also dug clay out from the wall to make a flatter, more stable slope. Our conversation takes us underground into the geological past. As Thiessen explains things to me, he draws a series of sketches starting with the glacial formations upon which we stand, including Lake Agassiz.

I am charmed that geological spacetime is a part of this man's everyday vigilance, a cognitive skill and material perception central to his professional work maintaining the right balance in the city's riverbank. Basically, a clay layer covers the whole region, and where it interfaces with the river, forming the channel walls, the force of the currents erodes it; large curves of clay slip down or, in the more vertical parts of the bank, fall off in big hunks. For his project, Thiessen (2010) studied the underground structure of the bank and then constructed three rows of columns, each 2.1 metres in diameter, and filled them with crushed limestone. The columns anchor the clay in the glacial till formation below.

2. *How the Technozone Negatively Affects Its Insides*
The Lyndale Project restores a bank that continues to confront forces both natural and technological. It is in part the result of MIT's macro-scale system-wide decision-making processes that causes bank failure on the Red as it passes through – even in the protected zone of the inner city.

To explain: the strength of the bank is affected by how much water infiltrates the ground from rain, snow melt, surface run-off, and flooding. Higher water pressure decreases soil strength and contributes to the instability of the riverbank. But bank failure tends to happen only after a flood. During a flood, the higher water pressure in the channel pushes back against the bank. But thereafter, the river level drops and sometimes it drops more precipitously than it ordinarily would because of MIT operations. When a sudden drop occurs in the pressure and height of river water pressing into the bank from the channel and does so at the same period as water pressure and water levels inside the bank continue to press outward, the possibility of bank failure increases.[6]

But MIT has responsibility for flood control at provincial scale. To achieve success, the operational tempo is keyed to the seasons.[7] Most relevant here are the routine water level adjustments in Shellmouth Reservoir, which regulates the volume of water coming into the system through the Assiniboine River and ultimately, at The Forks, into the Red River (see figure 3.1). In winter, MIT adjusts the water level in the whole system to keep it as low as possible (also relying on St Andrews Locks and Dam downstream north of the city). There are two reasons for keeping the river low. First, the lower the water in the channel, the more room the channel can accommodate spring floodwaters without overtopping the dikes. Second, the lower the water levels on the Red and the Assiniboine, the less ice forms in the channel.

Because spring hits the Dakotas in the south before it arrives in Manitoba in the north, the spring meltwater can rush into Winnipeg when the river there is still frozen. If thick enough, ice can block the meltwaters rushing in on the Red; in specific spots, it can force those meltwaters out of the channel and onto the flat floodplain. The lower the river level coming into winter, the less ice, and the lower the chance of having to break up and clear an ice jam. In short, to avoid ice problems, the river is kept low throughout the system.

But when MIT lowers the Assiniboine in the fall, it also lowers the Red along Lyndale Drive, which can trigger bank failure. (Although the Assiniboine flows into the Red and primarily keeps flowing north, there is a considerable backflow effect south of the confluence that can add to flooding in Norwood and Riverview.) So here is a case of scalar mismatch where operational decisions at the provincial scale adversely affect a riverbank that is part of the city's primary dike system. The many homes on the interior side of Lyndale Drive are set below the roadway-dike and are vulnerable to flooding should the dike system fail.

Geofluvial forces have always weakened the Red's banks. This is the nature of river dynamics. Because land beside the bank has always been subject to flooding, settlers have set their permanent dwellings well back from the edge. Eventually, the City claims jurisdiction of this piece of the floodplain, zoning it as public land. And since the Floodway, the neighbourhood has not been vulnerable to extreme flooding.

Here's the catch: operating the system that protects the homes from floods – lowering winter river levels in preparation for spring – exacerbates risk of bank collapse in the fall. In other words, water management techniques at the provincial scale contradict geophysical reality at the neighbourhood scale. Responsibility for correcting the negative consequences of system operation fall to city engineers like Thiessen, who – with a combination of public and private funds – tests and installs the set of rockfill columns that stabilize the bank.

This case shows that making and keeping terra dry, that is, turning a floodplain into a functional provincial territory, is an uneven and complicated achievement sustained by fixing whatever problems arise. Neighbourhood vulnerability can be characterized as a patchwork of good and bad. Because flood control engineering in the technozone is good overall, and because the unintentionally bad occurs on public land that is accessible and fixable, this whole arrangement is a rather mundane affair. In this case, territory is a political technology (Elden 2010) insofar as it constructs continuity of existing spatial arrangements according to settler-city standards. In this river reach, politicality becomes visible through an exchange between engineer and ethnographer.[8]

B. Riverview Neighbourhood, Red River's West Bank

On the opposite side of the river reach is another side to the story. As the outside bend of the Red flows toward The Forks, it erodes the forested shoreline of Riverview. But here, the city government does not have the authority to manage the bank failure problems that may be exacerbated by MIT's system-wide adjustment of river levels; it cannot restore the bank or try to prevent more serious bank failure because it does not own the riverbank.

> *Riverview scene (from field notes):* It is already mid-November, and I realize I did not know what cold was back in October. Thiessen meets me in Riverview to explain the riverbank predicament.[9] He shows me a section with a steep slope that has little wedge failures in it. These may be one, two, or several years old. We observe eroded sections where the bank has fallen and aggraded sections where the alluvium is piled high enough to strangle trees. The visual pattern indicates deep instabilities in the clay. When the bank does go, Thiessen predicts that it will "fail big."
>
> Recall that bank failure has happened recently along Lyndale Drive in Norwood, but there the bank is part of a park set aside to be used as public green space in recognition of its geophysical vulnerability. In Riverview, privately owned buildings and yards extend into the forested shore. The foundation of an adjacent apartment building has a crack going up the corner, while the cement platform serving as its parking lot had the soil eroded from under it. These signs indicate that the building has already been affected by the shifting riverbank.
>
> The dynamic geophysical processes that lead to a river's changing shape are natural, preceding all development. Neither city nor province can be blamed for causing or failing to prevent bank failures here on private land. But lack of liability does not obviate

the possibility of negative consequences, however. In any case, the geotechnical engineers cannot set about restoration; they can only watch and warn.

Private ownership of the riverbank thus complicates three political engineering problems:

First, risk: in the Lyndale Restoration Project, Thiessen knows that drilling into the rock to install the columns increases the risk of bank failure. He could nevertheless assume the calculated risk entailed in building-in long-term stability because the privately owned neighbourhood homes are some way distant on the other side of the dike. On the Riverview side, however, it would be difficult to ensure that restoration work would not have negative effects on the buildings.

Second, complexity: because the Riverview shoreline was built up by private property-owners who do not coordinate activities, shoreline fragmentation creates a level of complexity that makes stabilization efforts more difficult. For the city government, working on the Riverview shoreline is high-risk but low-priority; there are more feasible places to focus restoration efforts on.

Third, socio-economic considerations: bank failures do not coincide with property lines; they are much bigger. So, for example, if a problem section extends 150 to 200 metres along the shore and property-owners each own twenty-five to fifty metres of riverbank, the four to eight owners of properties at risk would have to cooperate in restoration efforts. Such cooperation is rare. Consider that the cost of a major work might account for a large proportion of a property's value, especially in larger, more desirable lots (about $10,000 CAD per lineal metre).

Property-owners are not prepared for and do not necessarily understand the risks entailed in owning shoreline. Imagine, for example, a hypothetical sixty-five-year-old pensioner uninterested in investing her remaining life savings to fix the portion of bank failure running through her lovely back garden. The way urban riverbank space is socially organized may lead rational decision-makers to opt for passivity instead of engineering. The socio-techno-natural contradictions are intractable, but they are probably not uncommon. Certainly, Thiessen does not see this problem being solved in his lifetime.[10]

City planners have divided Riverview's shoreline up into private lots. Architects, developers, and residents have built houses and apartment buildings too close to the riverbank. The provincial flood control system protects them from extreme floods, but the system-wide lowering of winter river levels creates the same problems of bank instability that threaten Lyndale Drive in Norwood on the other side. But in Riverview, no significant strip of public land separates buildings from river. Thus, the social and economic organization of infrastructural decision-making does not support adequate shoreline management. And even when the city creates a pilot grant program to try and overcome the neighbourhood's collective lack of capability, those whose properties are most in need of restoration do not apply for funds.

Thus, on the river's west bank, the structural balance of good and bad shifts toward the bad. Flood control engineering may be good overall, but because the unintentionally bad is occurring on private land that is inaccessible to city engineers, the bad cannot be corrected. Yet the situation is also mundane, seemingly ignored even by residents whose foundations are beginning to crack. The Riverview case shows that territory is a political technology problem insofar as it sets up conditions for collapse. Rather than the infrastructure-enabled sense of continuity observed on Lyndale Drive, the politicality of Riverview's spatial arrangement shifts to the near future, when impending collapse might topple privately owned buildings and parking lots.

Comparison of the two neighbourhoods dramatizes the fact that geo-culture is public culture. Collective action for the public good can be supported, as in Lyndale Drive, or inhibited or overwhelmed, as in Riverview. The resulting strengths and weaknesses can be read in the sociotopography and earthworks that characterize water-land juxtapositions in river cities.

Techno-zoning territory is never uniform and, although often mundane, is always political. The logistical power of flood control infrastructure imposes contradictory consequences. Even in the protected centre of a democratically governed technozone run by experts, mismatches trigger sacrifices. Such contradictions call for technical and symbolic fixes to sustain system legitimacy in geo-culture and provincial politics. Scholars and the media tend to focus on the most dramatic contradictions and injustices that arise when governments act to secure territory through infrastructure. I cannot help but wonder, however, how much of territorializing is more a combination of organized efforts and ad hoc muddling than the citizenry would like to think.

III. LESSONS FROM THE EXTREME FLOOD EVENT OF 1997: IMPROVISATIONAL EARTHWORKS AND THE FLOODWAY'S FIRST BIG TEST

Even when a disaster is by most accounts successfully contained, in some respects the river blasts past humanity's capacity to fully imagine, model, and (re)build environments.[11] When rivers interact with blizzards, torrential rains, high winds, and temperature extremes as did the Red and Assiniboine Rivers in spring 1997, they rise up to meet their geological destiny. When they rise, they challenge human assumptions. And all the careful data-gathering, synthesizing, and interpreting betray systemic vulnerabilities. Given geo-cultural throwntogetherness writ large and the particular spheres of unintended agencies activated in emergencies, this is no surprise. But the reverse is also true. The radical indeterminacies allow me to appreciate the momentous accomplishments of Manitoban engineers and political visionaries such as Premier Duff Roblin, who staked his political career on the Floodway ("Duff's Ditch"), all the more. Acknowledging their doubts and fallibilities, the impossibility of getting ahead of oncoming floods or of perfectly engineering the political aftermaths – these require courage, dedication, and flexibility. The flood of 1997 is the Floodway's first big test.

In the aftermath, Manitobans argue about the decisions and priorities, the successes and sacrifices. There are several official, scholarly, and popular accounts of events and subsequent debates about the kind of failures typical of any complex system.[12] In this brief recap of the 1997 flood, I focus first on the problem of entwined systems, specifically how transport and flood control systems depend on each other (e.g., primary dikes function as roads) and how their interdependence causes paradoxical effects. Like the contrived juxtaposition of railways and Lake Agassiz outburst pathways (in chapter 2), an image serves as a point of departure.

> The Flood of 1997 is approaching. Photojournalists and family photographers are everywhere building visual archives of flood memories for future generations. From the air, Ken Gigliotti

shoots a line of traffic "left with no place to go as floodwaters cut off Highway 75 outside of Morris" (in Currie and *Winnipeg Free Press* staff 1997, 27).[13] Tiny vehicles stranded on a narrow curve of terra are surrounded by the Red Sea. The 1997 photograph resonates with the old black and white photograph of the convoy on Norwood Bridge near The Forks taken right before the evacuation of 1950 (described in chapter 1). The camera's focus has shifted away from city centre, now in the protected zone, to the vulnerable region between Winnipeg and the US border. The uncanny repetition of lost landscapes repeats, yes. But also something else is worth noting. The viable waypoints in these lost landscapes are primarily built for transport, not flood control. A raised section of highway here, a bridge there, a lattice of raised linear structures have been transformed into safety zones. This secondary protective benefit derives from the entwined logistical power of transport and flood control infrastructure in the Red River Valley.

A contradictory aspect of the scene is less visible to most people. In the dominant flood imaginary of 1997, the Red Sea swallows the landscape of the everyday. It is most likely that those apprehending the scene are giving the Red River all the power, i.e., the agency of the river rising up out of the channel and on to the prairies is a natural disaster. What is hidden by the magnitude of Red River agency in the flood imaginary is the extent to which transport infrastructure causes crisscrossing overland flooding. In fact, during floods, roads and railway embankments can act paradoxically. A small selection from Simonovic and Carson's (2003, 353) detailed review of the 1997 flood in *Natural Hazards* illustrates:

> When the capacity of the channel is exceeded, the river overtops its banks and flows over the land. The extent of flooding depends on available water and the topography. In larger floods, the Red River can overtop adjacent roads and railway embankments. The flow then moves north controlled not only by the floodplain topography but also by roads and railways. These features confine the flow and sometimes act as obstructions to flow. Consequently, water elevations in overland flow areas can be higher than those in the adjacent main channel. The overland flow may then return to the main channel with destructive force by breaching embank-

ments. The 1997 flood resulted in a flooded area up to 40 km in width; many residents were flooded by overland flows. The nature of the overland flow is highly variable in both space and time; some tributary streams may flow in reverse as the flood wave moves down the Red. Sudden washouts of road and rail embankments and road cuts made by government personnel to reduce local water levels further complicate the picture. Models in use at the time of the 1997 flood were not capable of dealing with complex overland flows.

As Rannie (1999) argues in a North Dakota–Manitoba Academy of Science conference, the destructive impact of overland flows is further complicated by poor zoning of buildings and by drainage practices.[14] What a crazy mess to try and sort out! And sometimes, big things absolutely must be sorted out on the fly .,.

A. A System Design Glitch, the Z-dike, and the Gift of Blended Habitus

Probably the most dramatic, bigger-than-life story of the 1997 flood recorded in the book *A Red Sea Rising*, authored by Buzz Currie and the reporting staff of the *Winnipeg Free Press* (1997, 121), begins with Ron Richardson, a Highway Department hydrology engineer (55–8). The stage is set for suspense. The season's fourth and last blizzard starts 5 April, and rapid snowmelt follows. And this is just the beginning. By 23 April, plans for the greatest evacuation since the flood of 1950 is in the works (27). By 25 April, the whole system is maxing out. The Red is creeping up to the bridges near The Forks (109). By 27 April, the Floodway is full. I compose this story based on selections from the book:

> On 21 April, Ron Richardson is home studying a map he received from his former employer, Water Resources. He pays particular attention to flood projections in the area near his home. He realizes with dismay that the dike-wings on either side of the Floodway are not long enough, that the Red Sea might do an end run around them into the LaSalle River basin, an Assiniboine tributary.[15] If the Red flows into the LaSalle, it could circumvent the Floodway and pour into the city, forcing the evacuation of an estimated 100,000 additional people. Richardson calls his former col-

leagues, who talk to others. The expert team consulted includes MIT's engineer Steve Topping as well as old timers who have been around since the 1979 Red River flood. Pondering the situation, they realize to their horror that the Floodway has been built on the assumption of a PMF (probable maximum flood) of only 150 years. The forecast gives the team seventy-two hours to figure out how to design and build a Floodway defence for a flood with a much bigger PMF.

They bring in a private survey company to drive prospective routes in farmland near Brunkild. The survey team drives a truck with a rooftop receiving dish that bounces road elevations off a geostationary satellite at about fifteen miles per hour.[16] The team plots elevations and maps an optimal route along high ground. They accomplish in a single day what normally would take two weeks. Dike-building operations are turned over to Highway Department staff. People and machines work round the clock for three days to build a fifteen-mile, zigzag-shaped defence. The "Z-dike" is made of mud clawed from farmers' fields with backhoes, limestone fill from quarries driven onto the dike in thirty-six-ton trucks, sandbags as large as 3,000 pounds dropped into place with cranes. The deputy minister of highways also comes up with the idea of putting a breakwater in front of it (Currie and *Winnipeg Free Press* staff 1997, 62). His team drags or drives in enough derelict school buses and crushed cars to fill an eight-mile stretch.

Once the Z-dike is built, the question is, would the dike hold whatever may come? Officials are ready with an evacuation plan for the whole city, then 300,000 people. Warkentin, the river forecaster, is working day and night alone in the Water Resources building. His forecasts keep the Highway Department and Royal Canadian Mounted Police informed about which roads they can use and about if and where they might schedule trains through the flood zone.

Winds and rain, the big unknowns: There's a fifty-fifty chance that winds would blow from the south, whipping water into the Z-dike (ibid., 59). If winds from the south blow persistently against the dike, they would rile up wave action in the Red Sea that could "go through packed earth as if it were spun sugar" (60). But "[t]he gods smiled on the Brunkild dike builders." As it turns out, the roadways between the Red River and the dike slow down the flood's approach. And the winds blow in from the north, pushing floodwaters away from Winnipeg. Fortunately, too, the rains do not return. On 29 April, when the Red Sea laps up to the Brunkild gap, the Z-dike blocks its way. The Floodway holds, just. And the $100,000 CAD worth of sandbags a farmer buys saves his farm.

In retrospect, the Z-dike is hailed as a great accomplishment achieved in record time. But the throwntogetherness of its hasty construction fixes a vulnerability only to create another. Twenty years later, in 2017, a road sits on top of it, still ready to protect Winnipeg in the next big flood. But the farmers whose land lies in the dike construction zone feel that they have not yet been made whole. Their land's productivity remains diminished. They also wonder whether their livelihoods had to be harmed to the extent they have been, given that the 1997 floodwaters never did rise against the dike's full length. Perhaps, if the system design flaw had been caught earlier and there had been more time to plan, these farmers would not find themselves in the undesirable position of being ad hoc infrastructural outsiders.

But in geo-cultural terms, what stands out to me is the way one man, Ron Richardson, is able not only to pinpoint a vulnerability but also to inspire a huge landscape improvisation that averts a threat to the entire city. Richardson, as it happens, is a man benefitting from blended habitus, i.e., habitus that blends personal and professional dispositions and sensibilities. In the suspense of the upcoming flood, he is a homeowner assuring his neighbourhood's safety, a hydrologist who knows the science of water flows, and a government employee who is part of a professional network that can move his insight into collective action in record time.

The Z-dike and the many ring dikes, mounds, and sandbags tactically distributed across the landscape shore up the larger structures of the control system. It is fair to say that these supplemental earthworks are fine examples of the "creative urbanity" (Guano 2017) evident among Winnipeg inhabitants. That said, routine system operations can cause problems that these supplementary structures alone can't fix, even within the protected centre of the technozone.

IV. THE SECURITY OF TECHNICAL TRADITIONS

Clearly, insofar as it is practical, Manitoban experts have enough confidence in the system to keep trying to fix or work around common flaws and unexpected surprises that threaten viability.[17] But the flood of 1997 may be the last big flood that – even with all the crazy weather

– still fits into the traditional sense of systems and seasons. The Red Sea rose up in the spring with snowmelt; the Assiniboine, although active, played the usual secondary role. But the problem of overland flooding, its quixotic and unaccountable nature, as characterized in the technical reviews of Simonovic and Carson (2003) and Rannie (1999), offer a technical foreshadowing. When the floods of 2011 and 2014 come, the experts of seasonal forecasting and operations are taken aback by future possibilities they had not worried so much about before. The long and roguish Assiniboine would test the particular configuration of the twentieth-century system and how it negotiates with the flood bowl's throwntogetherness.[18] The thing about this metaphor, throwntogetherness, is its implication that even when forecasters, operations engineers, policy-makers, politicians, and householders miss their marks, they can adjust their timing, shore up the cyber-earthworks, and fill those sandbags and the coffee cups of volunteers and soldiers with exhausting alacrity.

Security = expertise + blended habitus + room to maneuver.

5

Furies of Wind and Wave (2011 Flood)

The Flood of 2011, which is estimated to have caused more than $1 billion CAD in damage, "may have been the most severe flood experienced in the history of Canada" (Blais, Greshuk, and Stadnyk 2016, 74). What measurement techniques are used to arrive at this figure, and how does severe flood experience unfold for different communities? Seeking answers, I discover a legally muted but topographically powerful internal frontier.

I. BALANCE AND LEVEL, IDEAL AND MEASURE

On calm days, lake waters quietly fill their beds, rising to unite shoreline inhabitants near and far. On calm days, their singular levels may inspire a sense of balance, of contemplation. Lake bodies offer a shared orientation and cultural coherence for those who gather beside them. When abstracted from the landscape, lake levels may also become a unit of measure, one that partakes of this sensory ideal even as it serves as a practical reference, a baseline comparator for stormy days. For experts, lake level indicates how much water a lake might hold, a data point collected among others to use in the quest to find system balance.

Water stewards set up gauges to track rising and falling levels. The numbers stream digitally, reflecting off lake surfaces up into satellites that beam them back down into office computer models (see figure 3.2). For engineers, lake level is the calculated balance between inflow and outflow, rainfall and evaporation. In the laws that support engineering, lake level provides a boundary and index range, a language for negotiating balance, and thus a technology through which the moving aquatic elements of lacustrine space can be governed. Lakes are an important feature of Canada's geography and geopolitics. More than half the world's total surface area occupied by lakes is within Canada (Strahler and Strahler 2005, 479); these lakes hold about 20 per cent of the world's fresh water (Pearce, Bertrand, and MacLaren 1985, cited in Linton 2010, 292, note 5). Manitoba thus seems a great place to study human–lake relationships. As waters assemble themselves in and around these phase-shifting (solid/slush/liquid/vapour) and shape-shifting bodies, lake levels organize thinking and action in the social production of space (Lefebvre 1991).[1]

Water's agency as a social actor, I argue, cannot be extracted from the shapes it assumes without loss of ethnographic understanding.[2] Lakes as actors play roles different from those of rivers, even though the same elemental water may flow through both, even when extreme floods blow their normal shapes out of proportion.[3] In the time of impending flood, the usefulness of lake level as organizing concept in flood control becomes most crucial and explicit.[4] A lake's shape expands and contracts with its level. Lake-level measurements guide frantic communal efforts to sandbag narrowing interfaces between homes and waters; they time evacuation alerts and key emotions to immanent events. In the time of impending flood, attunement to lake levels is a floodway tradition, a sensibility that arises with tangible measures of feeling and thought, a key element in an emergent-yet-familiar "structure of feeling" (Williams 1977, 128–35).[5]

Lakes and the ideals they inspire have been incorporated into a hardscaped sphere of influence, the flood control technozone under MIT's jurisdiction (see figure AII.1 in Appendix II). By the 1960s and 1970s, regulating the levels of Lake Manitoba and Lake St Martin became central to the social-spatial distribution of protection and hazard. Opening and closing inflow-outflow structures on lakes is a routine feature of risk management. And for good reasons, they may be subject to public contestation, especially in extreme floods when states of emergencies free provincial decision-makers from normal operational rules and guidelines. Engineers may stretch lacustrine water storage functions to, or even past, their limits, intensifying flood hazards for those who dwell in the system's edges. Because, as the mantra goes, floodwaters will go somewhere; when diverted into lakes, rising levels reveal the desperate, semi-accountable edges of experts' protective intentions.

II. WHAT HAPPENED TO INFRASTRUCTURAL OUTSIDERS AFTER THE FLOOD OF 2011?

A. Yellow Ribbons in a Landscape Flooded Too Long (Ethnographic Scene 1, South)

Everything depends on where you are in the basin.
Scott Forbes, biologist and cottager, Lake Manitoba's south end

One of the yellow ribbons is tied to a roadside telephone pole at the roadside edge of a neighbour's yard, officially indicating how high off the ground cottagers are supposed to build the ground floors of yet-to-be-rebuilt houses. If they rebuild. But will they, given the uncertain future of such an endeavour? Beyond, on the lakeside edge of the lot, the white super sandbags sit beside gigantic black geotextile tubes (geotubes). Their persistent, ubiquitous presence belie the serenity of the scene they guard on this August day, three years and sixty-four days after an intensely localized event within a major flood disaster occurred. Although progress has been made – for example, the road has been rebuilt, and wetland restoration is proceeding apace – neither of the two main hazards, one natural, one infrastructural, that set up Twin Lakes Beach for the 2011 flood disaster has been changed. The community's location puts it on the receiving end of northwest winds crossing the long, shallow length of Lake Manitoba, whipping up and gathering waves before slamming them into the cottage-lined isthmus between Lake Manitoba, the province's third largest lake, and the little, marshy Lake Francis. Nearby, the Portage Diversion dumps in water from the Assiniboine, diverting excess before it threatens Winnipeg downstream, keeping lake levels higher than they would otherwise be (see figure 1.2). There's no sign that MIT will ever release Lake Manitoba from this basic storage function in flood control operations.

The source of the account in this section is Scott Forbes, a (settler) cottager who owns a now mostly soggy stretch of beachfront property. Because he is a behavioural ecologist at the University of Winnipeg who does research on marsh-nesting birds and on fish, his neighbourhood perspective is uniquely informed by his scientific knowledge.[6] Passionate about the vulnerabilities that flood control imposes on infrastructural outsiders, he contributes regularly to the letters to editor section of the *Winnipeg Free Press*. His letters focus particularly on what he considers the chronic misuse of the Portage Diversion. The disturbing reality of provincial law, Forbes tells me, is that once the province declares a state of emergency, "they can do whatever they please."[7]

In the letters, Forbes crafts angry yet persuasive public policy arguments that vary with flood events, engineering solutions in play, and accompanying provincial politics related to decision-making. His core argument is that the diversion should only be used in extreme circumstances, not, as MIT has been doing, as a chronic management solution for preventing wet sidewalks down at The Forks in central Winnipeg.[8] As we ride the 100 km or so from the city to the lake in his truck for this interview and site visit, we talk about how the convergence of a storm and engineering practice turned the flood of 2011 into a disaster for his little community of Twin Lakes Beach.

That spring, the Assiniboine had already been swollen, and engineers had been diverting the excess into Lake Manitoba for some time. But it wasn't until 31 May that disaster hit. The storm blew seventy km/hour winds, which caused the lake level to rise an extra foot. The strong winds turned the already over-full Lake Manitoba into a raging inland ocean. Water rose two feet above the high ground of the road and then flowed inland for two kilometres, "as far as the eye could see."[9] The windstorm also drove water from Lake Manitoba into ditches and, via ditches, overland into Lake Francis; many houses were more affected by overland flooding from ditches than from Lake Manitoba itself. Unlike wind, overland flows (or overflows) become part of the calculation discerning the difference between natural (unregulated) and artificial (regulated) flows for the purposes of allocating recovery funds.[10]

The 31 May storm obliterated most of the property on the east side of the lake. Most structural damage happened then. In contrast, most ecological damage resulted from the subsequent long time span in which water remained on the land. Thus, structural, ecological, and social hazards unfold in different "timescapes," as do recoveries (cf. Adam 1998). In any case, it is not unlikely that some similar constellation of damaging hydrological forces and infrastructural conditions could assemble again.

Lake Scene 1 (from field notes): We arrive and sit at a picnic table eating lunch, watching the redwing blackbirds glide in and out of the lakeside scene. Forbes describes how the post-flood landscape has been slowly returning to a different version of its former self. The lake returns to its normal operating range a little over a year later, in the summer of 2012, but nothing else has come back yet; in 2013, the thistles and dandelions grow eight feet tall aided by all the new nutrients in the soil; now in the summer of 2014, swaths of invasive purple loosestrife push into meadow grasses. Only a few scraggly cottonwood trees survive, testifying to the riparian forest edge that once served as protection from the harshest winds (at least on this and other homesteads where trees weren't cut down to create better lake views). Except for the road, rebuilt to higher ground, the edge between lake and land, or wetlands and dry land, is ambiguous. Water creeps toward homesteads, an implacable threat to security.

Without wading into the wetter parts of the landscape, I can't see the carp population that Forbes says has exploded post-flood. The carp overpowered the fish gates of the Delta Marsh Project that had attempted to exclude them from the coastal wetlands, which are once again well packed with the water-filtering rhizomes of bulrushes (cattails).[11] As the carp females burrow into the mud, as they are wont to do, they resuspend stored phosphates, contributing to the murkiness of the water and to eutrophication. Pike lurk in the shallows with the cannibalistic carp, fish that eat their young. The 2011 flood was a gift to both these aggressively invasive species, which between them are causing a severe loss of biodiversity. In contrast to these hidden underwater events, everywhere I look I see the leopard frogs Forbes points out, hopping, plopping, or pausing with faraway looks in their boggy black eyes. In the moment, it is hard for me not to welcome the quiet guttural sounds of this herpetological invasion.

A little cottage, the one Forbes and his wife and sons built together to provide shelter while building their big house, is all that remains. Forbes notes sadly that his family fares better than most. Super sandbags and geotubes (state-of-the-art flood control materials for wherever there is wave impact) sit around the property next door; the boarded-up windows of this four-season house are evidence of its uselessness as shelter (see figure 5.1). The couple who lived there before the storm, one eighty, one seventy-five years of age, now have to stay in the city. On that 31 May, water

5.1 A boarded-up house on Lake Manitoba's south shore, three years after the 2011 flood. Beside it, white super-sandbags and black geotubes await the next disaster.

had swamped the home, forcing the wife to jump out of the bedroom window onto some sandbags so that she could make it through the wind to the truck before the road became impassable. (They hadn't realized at the time that the regular-size sandbags are useful on rivers but not lakes; barriers built with smaller ones can't withstand the waves.)

The storm wrought random havoc. The elderly couple next door to Forbes returns afterwards to find their pot of coffee undisturbed in the kitchen while the cottage on their other side was completely obliterated. Indeed, all along the isthmus road, vacant lots stretch between rare inhabited ones. Some cottagers face so much aggravation and difficulty they decide not to rebuild. Much debris has already been cleaned up, tree limbs and branches dragged and dumped into piles, the stumps ground or left to rot. Roadside signs advertise businesses such as construction and excavating, tree removal, electrical repair, eaves troughing, and Internet. But Gratton's General Store, the community meeting place and source of Chester's Fried Chicken, still has a big handwritten CLOSED sign taped to the inside glass of the door.

Giant backhoes work at tearing down houses, at times inadvertently counterproductive. For example, talking on his cellphone while driving his truck with the bed turned up, a driver hadn't realized that he was snapping telephone poles like toothpicks along the way. Manitoba Hydro wants $2,000 CAD from the householders to reconnect their houses to the electric lines, even though the trucker was the one who caused the problem. Two friends of Forbes get fed up after this happens and decide not to rebuild. The province allocated $65 million for a program to flood-proof properties, but even if they had the choice to rebuild, no one is going to come back as long as they know that the vulnerabilities caused by MIT's continued reliance on the Portage Diversion persist. In fact, folks who live at the other end of the road keep their suitcases packed just in case a flood should occur.

In this wind-blasted geography, the yellow ribbons of the law tied to utility poles index both obstacles and promises. Rebuilding a home not at the level it had been but, instead, much higher off the ground takes considerable fortitude and money. Permanent residents, defined as those who live on the lake year-round, can rely on recovery funds from the province should they wish to rebuild. Provincial financial assistance is not available to the majority, however, who are seasonal residents. Most cottagers, like the Forbes family, do not endure the winters by the lake but rather have permanent homes in the city. Many built their cottages in expectation of retirement.

But the economy of St Laurent and Woodlands, the two rural municipalities in which Twin Lakes Beach sits, home to many Métis and Anglo-European farmers and ranchers and others, largely depends on seasonal inhabitants. So the province's recovery efforts fall far short of those needed by communities that rely on income-generating warm weather economic activity by the many to sustain winter life for the few. In this sense, the legal division between seasonal and permanent cottages, though not without logic, inserts another kind of random disruption into recovery efforts. Like the yellow ribbons, index of obstacle and promise, the effects are mixed.[12]

As I write, I recall now what it feels like to arrive at Forbes's homestead in the south basin and view the scene from the picnic table. In my mind and in my photographs, I see a beautiful lakeside, swathed in the bright green of grasses and flowers, birds on the wing, frogs scattered amphibiously. Were it not for Forbes's translation and walking tour, I might have been content to enjoy the beauty of that sunny day. I would never have been able to apprehend and communicate the Twin Beach Lake community's transition into and out of a landscape of loss.

B. Flood Control Chaos in Lake St Martin
(Ethnographic Scene 2 North)

Separating land from water on the earth's surface is one of the most fundamental and enduring acts in the understanding and design of human habitation ... Is this separation found in nature or does nature follow from its assertion?

Da Cunha (2019, 9) in *The Invention of Rivers*

In its northern end, Lake Manitoba has one outlet – Lake St Martin. While the province has appropriated all of Lake Manitoba's long extent for storing diverted floodwaters, the impact on the northern end has been qualitatively more devastating. It is the raw edge of MIT's technozone. The community known as Lake St Martin First Nation was completely destroyed in the flood of 2011. The decision-makers knew that that the operational decisions they were making that spring *might possibly* have negative effects on Twin Lakes Beach. But they must have also known that they would *most likely* have dire effects on the Lake St Martin communities.[13]

By the time I do fieldwork in 2014, Lake St Martin evacuees have no recourse but to try and survive for years in government-provided one-per-family-group hotel rooms. Their suffering is on regular display in the *Winnipeg Free Press*, known to all. But for the most part, my study focuses on the knowledge and practices of flood control experts and to a lesser extent on flood memories of ordinary inhabitants in and near Winnipeg, the latter a topic already given great attention. In the few instances where happenstance makes possible a conversation about flooding and flood control with a First Nation person, I follow the trail until it disappears. Like most Winnipeggers, potential contacts are not anxious to join conversations about the sadness of flooding with a stranger who is unlikely to make a tangible difference in their lives. And in the field, I don't push it. After 2014, once fieldwork's logistic constraints are behind me, I live Manitoba in thinking and writing mode. Building on some data from participant observation and my archive of secondary sources, I bring edged-out pieces of the flood control puzzle into play. But I am nowhere near this phase of the project when I am confronted with the pain my detachment causes another:

> One day, 5 November, I give a lecture at the Winnipeg Public Library entitled "Flood Control and Environmental Justice in the Manitoban Prairies." The audience is a mix of interested people from various walks of life, including scholars and experts, some of whom I had interviewed. I present some of my preliminary analysis of how the geological past animates contemporary urban flooding and flood control. I weave in themes of law, engineering, and landscape, drawing on visuals, such as a photograph of the post-2011 flood scene with yellow ribbon. Among the questions and comments that follow my lecture, a natural resource scientist originally from Lake St Martin First Nation, Dr Myrle Ballard, asks how it is possible for me to leave this community out of the talk. Wow.

She is totally right. How, indeed? We meet afterwards for an informal lunch-interview at my house.[14] We plan to drive together up north for a site visit, but the ice moves in before we have a chance. Ballard has given me a DVD copy of the 2012 documentary film she directed and co-wrote, *Flooding Hope: The Lake St. Martin First Nation Story*:[15]

Standing at the foot of a map of loss is clarity.
    Leanne Betasamosake Simpson (2017b, 15) in *As We Have Always Done*

> *Lake Scene 2 (from Ballard et al.'s documentary* Flooding Hope*):*
> The audience revisits Lake St Martin after the flood of 2011 with Ballard as she returns with a group of elders and a film crew to retrieve belongings that unlock personal memories such as family photographs. The camera shifts back and forth from the wasteland surround to the women elders speaking on their porches as they might have done in ordinary times. Photos and video of hap-

pier times are spliced in and juxtaposed with those of infrastructural crimes (e.g., Fairford River Control structure, bulldozers on the Emergency Canal). Even a complete stranger could not mistake loss for beauty here, where culture meets the limits of flood control.

The frogs are not staring from the reeds and roadside but instead are living in what used to be your house, with mice that have eaten holes in all your clothes. But could this be your house with groundwater lapping right under the floorboards? Where black and yellow mold swarms slowly across the walls, like in Spike Lee's 2012 film *When the Levees Broke* about the ninth ward in New Orleans after Hurricane Katrina? Where temporary trailers become permanent hellholes and the government builds a new school on a snake pit? Really? Where thick clayey gumbo smothers everything between porch and horizon. The lake and creek waters flow so far from the porch's visual reach, they seem as distant as are the vanished people once fishing, hunting, dancing, gathering plants, doing bead and leather work, studying, cooking, making music. The healthy lake and creek waters are as distant as the scenes that the now faraway children draw of all this once-upon-a-time, a sun shining brightly over their simple stick figures. As distant as one's family and community scattered across the province or stuck for years, depressed, overwrought, suicidal even. Split among crowded Winnipeg hotel rooms where their bodies are meant to be sustained by insufficient food stamps, where the Anishinaabe elders' language, Anishinaabemowin, becomes foreign, and where their spirit languishes, waiting for the day they can return.

Dispossession becomes a default process that layers along with race, gender, and class hierarchies, sedimenting inequality in the province and in North America more widely (Gill 2000; Coulthard 2014; Hogue 2015; Simpson 2017a and 2017b; Razack 2016). The Manitoba flood of 2011 clearly shows how official decision-making benefits from colonial practice. Recall the nineteenth-century provincial maps that stabilize the habitus of settler land ownership (see chapter 1). Following Da Cunha in the above quote, these and later maps encode and enact the reconfiguration of settled nature, of rivers as entities separate from land; seemingly by default, they unequally apportion the pure havoc of this rearranged nature to Indigenous outsiders. In other words, colonial settler habitus blends passively into contemporary infrastructural decision-making as a necessary given rather than an intention. As details emerge from the technical recesses of expert discourse, I find that it is possible to analyze the paradoxically central role of distant First Nation lake-dwellers in Winnipeg's flood control system from a different angle.

III. INFRASTRUCTURAL REGULATION: CALCULATING NATURE AND ARTIFICE

What imaginative work do measurements do?
>Marilyn Strathern (1999, 221) in *Property, Substance and Effect*[16]

The regulatory standards useful for auditing system failures and negative side-effects have been designed to sort out specific events as fairly as possible. They are not primarily designed to cultivate healthy webs of ecological or cultural relationships, although these twenty-first-century intentions are to some degree factored in now.[17] In the aftermath of disaster, the law relies heavily on lake-level calculations by engineers to sort out questions of fairness and restitution. Sorting and modelling blame (for what happened) and guilt (for who is responsible) in the unequally distributed impacts of lake-level management – dispossession or devaluation of homes, lands, livelihoods, roads, communities – must have a comparator.[18] The harm caused by flood control structures is measured against what would have occurred had they never been built.

In other words, engineers engage in a studied attempt to calculate the nature that precedes their artifice. Interpreting this effort culturally, one could say that by reaching for a calculation that captures a sense of an original, the inter-glacial flood bowl into which settlers sailed rather than the one made by engineers "provides the comfort of an existential foundation" (Bennett and Chaloupka 1993, ix).[19] For engineers, the calculation is pragmatic but not uncomplicated. Sorting and modelling of blame and guilt depends on the calculation of nature's more predictable aspects while leaving out nature's furies of wind and wave. This is accomplished through quantifiable artifacts such as the "wind-effect-eliminated" water levels (WEE) used to compare different scenarios (MIT 2013b, 100–5). The WEE calculation for Lake Manitoba is based on readings of two gauges, one on the north end of the lake and one on the south*west* end, a procedure that "smooths out most of wind effects (61)." Curious tactic. Because prevailing winds are northwest, storms, like the one in question on 31 May 2011 that smashed into Twin Lakes Beach, would be best considered by a gauge in the south*east* basin.

Is this a bias in the calculation that favours MIT's ability to keep lake levels higher than is fair? And anyway, sometimes winds are completely still, sometimes they wreak havoc, so what function does an average have other than to leave wind out of the sphere of calculation and responsibility? To be practical, from an operator's perspective, wind dips well into an effect that arises in a sphere of unintended agencies. Fickle, wind is difficult to predict with any spacetime precision, yet it can be utterly consequential.[20] But one thing is incontrovertible – the higher MIT maintains the lake level, the greater the flood risk from strong winds. One could say that the WEE calculation ignores the precautionary principle. That is, when you're not sure what might happen, err on the side of caution. In this case, if the precautionary principle were put into effect for *lake residents*, MIT would probably be required to maintain lake levels low enough to account for wind's worst. Perhaps the precautionary principle is in effect but for the benefit of *city-dwellers*. The underlying social inequity is disguised by the apparent neutrality of the WEE calculation. In this sense, the WEE calculation helps law to create a wind-blown atmosphere of fairness (cf. Philippopoulos-Mihalopoulos 2015, 107–50).

Winds aside, MIT's calculated difference between nature and artifice is a mechanism that methodically condenses and transports hydrogeological data into law. Despite its empirically grounded logic, however, I find the nature that emerges in the intersection of engineering and law a queer actor indeed. Geoscientific knowledge splits into twin forms – an original, prehistorical hydrological scenario splits off from an invented one (cf. Strathern 2005, 39, cited in Lebner 2017, 17). And this is only one categorical cut among many that assessors enact in the quest for assigning blame and guilt for post-flood damage. Through such selective processes of measurement repeated systematically, ritualistically even, across lake-spaces in different floods, the state performs public accountability.

Much culture-work entails coming to terms with the geophysics of lacustrine excess and provocation. Politically, ideals of fairness – evaluating the degree to which decision-makers exercise balanced judgment when performing engineering prowess – tend to escape the practical confines of official measuring efforts. In times of relative equilibrium, the day-to-day work of staff engineers is overall system management. How they move water through the system within acceptable parameters is guided by data and set by law. After a flood event, legal formulas for auditing state responsibility for harm can guide discussion but rarely end debate. Infrastructural majorities who benefit from provincial protection will appreciate the pragmatic rationale for the reliance on lake levels in engineering and law; infrastructural outsiders, by contrast, will dispute the human arrangement of water flows that create the levels that disadvantage them in crises. They may dispute the fairness of using technical artifacts like WEEs that effectively eliminate wind damage from assessments or dispute categories of eligibility that include and exclude groups from timely, or any, distribution of recovery funds. The tension between majorities and infrastructural outsiders, and among outsiders whose geographical or sociolegal position in infrastructural systems may pit them against one another, is a given feature of political engineering in the technozone.

During this analysis, I find myself pivoting between lake balance and lake level as framing concepts. The balance frame captures this sense of things: Lakes are always in flux and always in search of balance. The search for balance, a basic feature of lacustrine geophysical nature, manifests and practically orients a poetic ideal of balance at the heart, not just of flood control but of social and environmental justice and Indigenous tradition more broadly.[21] The level frame captures another: The levelling action that lakes perform moves into legal, engineering, and popular floodways in curiously specific ways. Argument, calculation, and speculation following the flood revolve around the management and justification of physical lake levels primarily in Lake Manitoba and Lake St Martin, which are gravely affected by the Portage Diversion of the Assiniboine's prolonged excesses.

I could rely on a simple engineering trick of scale to resolve my framing dilemma. In the matter of lake flooding, hydrological balance pertains to the whole riverine network. Lakes are an integral part of the whole. Engineers use a Water Balance Methodology based on the physics principle of the Conservation of Mass to calculate the multitude of unregulated, pre-settler flows moving through interlinked water bodies in the system (MIT 2013b, 60). In this schema, a functional balance in the whole system depends on the sum of levels in all component water bodies. In other words, engineers might seek balance in the whole system by forcing any particular lake level component to be completely out of whack. The engineering logic of a balanced whole (the ideal) made of out-of-balanced parts (the unfortunate reality) enters directly into law, forming the scientific evidence for the calculation of (post?)colonial justice. If an out-of-whack flood level in a particular lake is traced to engineering operations, then the government owes recompense to those negatively affected. Once a flood disaster is sorted, the government can mandate additional requirements from lake-edge dwellers if they want to stay where they are in the future.[22] The yellow ribbon mandating the standard level at which houses must be rebuilt is one such requirement at Twin Lakes Beach, where habitation continues to be possible.

For infrastructural outsiders who are settler minority citizens, this logic is workable and leans toward fairness. Yet the models and empirical data-based calculations that go into them make assumptions. They also leave out a great deal of complexity. The sources of distortion include arbitrariness in setting the model's boundary conditions in spacetime, difficulties of measuring lake inflows from agricultural fields and ditches and of calibrating overbank flow conditions, and paucity of data about lakes and rivers at the northern edge of the system (MIT 2013b, 63, 94–5).[23] What data get left out of collection and analysis, by what process does data get included, and with what consequences, for whom? Myriad assumptions get made; some of them will be consequential when actual events unfold, and most of them will be invisible to the public. For example, in the calculation of WEEs as sketched above, the consequential fact that the southeast basin is differently affected by wind is erased from the average and hence from modelled or calculated bases for post-flood decision-making. This greatly affects Twin Lakes Beach.

Complexity always produces ambiguities in data collection, modelling, and interpretation (Odoni and Lane 2010). Engineers recognize them in their technical reviews as caveats. Articulating caveats performs transparency. After all, this is where I get the information in this section. Nevertheless, when push comes to shove, the caveats drop out of crisis decision-making and retrospective assessments. Put otherwise, given that geophysics is only partially calculable and that an "original" or pre-settler topography can only be estimated, how do engineers actually navigate between uncertainty and policy requirements?[24]

Recall the shape-shifting form of Lake Agassiz that geologists recorded in the post-outburst sediments of the contemporary flood bowl (in chapter 2). In the reconstruction of dynamic scenarios, how do geoscientists decide which moment is the best one to specify as the original? Sorting out the answer to a question like this depends on logistics as much as on scientific method. In the end, the process has to entail some collective agreement to go with a best guess (backed-up with lots of redundancy and replication). Authors of the official technical review, at least the one for 2011, which I study in detail, delineate several different forms of incalculability that make their data fuzzy. Yet, in the end, data fuzziness is set aside, and law and policy decisions are made and deemed solid. This leaves a lot of wiggle room to produce and contest geoscience-based rationales at the interface of law and engineering.

Alas, as the ambiguities fall away, the guiding result tends to support the status quo even when ineffective. And making things even more difficult for dwellers on out-of-whack lakes, First Nation territorial sovereignty exceptions add jurisdictional ambiguities that increase the inertia triggered by relying on engineering and law to calculating balance (as remarked upon in Ballard's film). Responding to Strathern's question quoted above, responsibly reproducing inertia seems to be the imaginative work that measurements do.

Ethnographically defined, the floodways of blended habitus include the complexities and the indeterminacies of experience. So I keep both level and balance in the analytic frame as empirical measure, calculation, and model and as stories and ideals. Initially, the goal of my analysis of the flood of 2011 is to delineate productive ambiguities that might help to enhance the pursuit of fairness for infrastructural outsiders or, at least, to provide some cultural ground for decoding the spaces of possibility that separate contradictory perspectives. As my analysis of 2011 starts to gel, I discover that the tactical uses of level and balance in the discourse of engineering and law are important for understanding currently accepted obstacles to social and environmental justice.

IV. BEFORE THE FLOOD: GEOSCIENCE AND ENGINEERING INTERVENTIONS

A. Deep Time Connections between the Assiniboine and Lake Manitoba

Lake Manitoba and the Assiniboine have had an on-again-off-again relationship through the Holocene Epoch (see chapters 2 and 8). Even before Lake Agassiz is born of glacial meltwaters, the Assiniboine runs eastward toward the continental interior from what is now Saskatchewan. With glacial meltwaters expanding its power, the Assiniboine's spillway creates a large valley (and floodplain). When the surging river runs down off the upper prairie and into what would become Lake Agassiz, the river acts like rivers do when they encounter sea-size water bodies – it makes a delta out of the glaciofluvial sediments that it has gathered en route. Over the thousands of years of running through the downward tilting delta of its co-creation, the Assiniboine cuts a narrow channel for its own passage; eventually, the delta that confines it ends, and the Assiniboine drops down into the flat floor of Lake Agassiz (now the Red River floodplain). The slope of its bed flattens out, slowing its velocity, decreasing its capacity to carry sediments, which it deposits on what we now know of as the fertile and flood-prone Red River Plain. Thousands more years of making and running through the mix of soft and gravelly sediments first produced in the Assiniboine–Lake Agassiz encounter, the river initiates several different pathways, moving in and out of the various paleochannels of its own creation through what becomes a prairie terrain that is flatter and lower than the steep-sided channel walls of its origins. In short, portions of the Assiniboine riverbed are perched above the surrounding land (MIT 2013b 1, 15).[25]

Lake Agassiz eventually disappears in the Outburst Quartet, leaving remnants of itself across the prairie. In what Rannie et al. (1989) call Assiniboine's first phase, Lake Manitoba is the meltwater remnant in closest relation. For a time, there was a low network of channels through which excess water flowed bi-directionally from one to the other. When Lake Manitoba is high, it becomes a tributary of the Assiniboine; when the Assiniboine is high, it sends its floodwaters to Lake Manitoba. The Assiniboine eventually emerges from the delta into an alluvial fan to make an array of other channels (see chapter 8). But then, sometime in the middle of the Holocene, when historical time first intersected with geological time in North America, give or take a few thousand years, the earth's slowly changing orbit changed enough to trigger a period of increased solar radiation and, in turn, an unusual warm and dry period in the middle latitudes. Lake Manitoba dries up. And if it hadn't been for the Assiniboine, which returns for a spell and shares some of its freshwater flow with the dry but awaiting lakebed, the south end of what is now Lake Manitoba would be part of the prairie, inhabited by canola and dairy farmers perhaps.

We know this portion of the Lake Manitoba–Assiniboine relationship from research on the opaline-glass bodies of diatoms, whose diversity of form and niche and widespread distribution in lakes and rivers help geoscientists infer the track of the Assiniboine as it strays toward and away from Lake Manitoba and closer to Lake Winnipeg in its second phase within Holocenic spacetime (see chapter 6).

B. Reengineered River Basins

The space of capitalist accumulation thus gradually came to life, and began to be fitted out.
Henri Lefebvre (1991, 275) in The Production of Space

[T]he very exercise of transparency informs the way that certain processes and events, such as environmental and social impacts, become the objects of public dispute, while others do not.
Andrew Barry (2013, 17) in Material Politics

When cities and their twentieth-century flood control engineers begin eyeing the hydrogeological relationship between river and lake, the Assiniboine has long since parted ways from Lake Manitoba. Nevertheless, that low networked topography of Holocene's yore leaves evidence to find and exploit. They built the Portage Diversion into this low topography, digging out and hardening a channel so that it works only in one direction, south-to-north, allowing engineers to protect Winnipeg (primarily) from Assiniboine flooding by diverting excess water into Lake Manitoba and from there to Lake St Martin. Thus, engineering mimics and builds upon while hardening and simplifying the ancient two-way flow relationship between lake and river. In its search for hydrological balance at the provincial scale, engineers alter relationships among the natural water bodies it mimics and clads in concrete.

As Blais et al. (2016, 77) point out in their study of the 2011 flood event, the Assiniboine River Basin has been independent of Lake Manitoba for thousands of years, but the operation of the Portage Diversion effectively makes Lake Manitoba and its outlet into Lake St Martin part of the Assiniboine Basin. Instead of the Red being the Assiniboine's only outlet, the lakes now act as secondary outlets. The rearrangement dramatically transforms the size and shape of the Assiniboine Basin on a spatial scale that almost rivals what Manitoba Hydro (2011) does by rerouting the Churchill River system into the step dams generating electricity in the Nelson River system. Indeed, the expansion of both the Assiniboine and Nelson River basins divert excess waters into the hydroelectric dams stepping toward Hudson Bay (see figure AII.1).

In effect, the logic of provincial scale made manifest by the construction and operation of the Portage Diversion treats the hydrogeological configuration of prairie rivers and lakes, as known since time immemorial, as arbitrary. Engineers leapfrog over geological time to reconnect the Assiniboine and Lake Manitoba. They do not attempt to mimic the complex flows of bi-directional networked channels. Instead, engineers stop spacetime in a singular configuration-moment to build a one-way, rudimentary, digitally operated cement channel. And thus, the law, the legislated texts of which set engineering in motion, picks and chooses among geological precedents to find the singular pathway that advances urban interests.

Even before the construction of the Portage Diversion, another more northern yet major flood control structure alters the relationship between Lake Manitoba's northern end and Lake St Martin, which are connected by the Fairford River. Indeed, the Fairford River moves into the sights of engineers soon after the province's formation. In his *Geological and Natural History Survey of Canada of 1888–89*, J.B. Tyrrell (1890, 19–21A, cited in Upham 1895, 61) refers to what I interpret to be two descending stretches of the Fairford River, one stretch between Lakes Manitoba and St Martin and another stretch, now called the Dauphin River, between Lakes St Martin and Winnipeg. In the flat prairielands, it is no surprise that these descents attract early settler attention. It would not be long before engineers would borrow the combined logistical power of water and gravity.

Settler-engineers first target the Fairford River following a period of flooding and high water in the late 1890s. A new Department of Public Works builds a canal between the Fairford and Lake Manitoba to improve outflow and provide relief to settlers. It doesn't work. Later, in the *dry* period of the 1920s and early 1930s, the province builds a control structure on the Fairford River to keep the water *in* Lake Manitoba. In 1961, when Manitoba cycles back into a wet climate, engineers again reverse the design of the Fairford River Control Structure, this time to facilitate outflow. Engineers base the design capacity of the Fairford River Control Structure on historical flows along the Assiniboine. They project estimates of what they believe they will have to send through the Portage Diversion once it is built and put into operation in 1970.[26] In other words, how much water the Fairford River Control Structure is capable of moving is based on a rough mid-twentieth-century guess. Of course, now with climate changes wreaking havoc with historical flow patterns, this structure's design appears to be even more untenable.

Before all these interventions at the north end of Lake Manitoba, the two lakes exhibited a positive relationship, i.e., over the long term, their lake levels would rise and fall together. Higher Lake Manitoba levels created higher Lake St Martin levels; higher levels in both lakes caused more water to flow out of St Martin toward Lake Winnipeg through the Dauphin River. But the 1961 control structure decouples this balanced relationship. After regulation, Lake Manitoba's water level could be below flood stage when Lake St Martin is above flood stage. Why? Because artificially high outflows from Lake Manitoba are heading through the Fairford River Control Structure into Lake St Martin. Not only are the levels of the two lakes decoupled, but the water levels in Lake St Martin are too high more often and fluctuate more frequently. For many years, living at the north end of the flood control system made the lives of the people in the First Nation communities on Lake St Martin miserable. But the Flood of 2011 makes living there impossible.

C. Reconstructing the Lead up to Disaster: Actions of Rivers, Lakes, and MIT

The flow of water down the Portage Diversion channel in 2011 was unprecedented. A record was set for the longest period of operation in one year at 126 days, diverting a record volume of 4.73 million acre-feet (5.83 million dam^3)[27] of water.

MIT (2013b, 77) in 2011 Flood: Technical Review of Lake Manitoba, Lake St. Martin and Assiniboine River Water Levels

[A]t the time of the May 31 storm event, there was no artificial flooding. In fact, on May 31, Lake Manitoba was approximately 0.5 feet (15 cm) lower than it would have been under unregulated conditions.

Ibid., 110

How the authors of the technical review move from the first statement above in section 4.0 that describes flood operations in 2011 through a series of calculated explanations illustrated by tables and graphs to section 6.2 in "Conclusions and Next Steps" is quite masterful. Between the two statements, they perform an earnest effort to be fair and comprehensive, detailing a series of assumptions, caveats, and uncertainties that might throw quantitative assessments off. Then, in the end, they let the problems fade into the background. Is this honesty-without-follow-through simply the way that science must perform transparency and translate geoscience and engineering into policy? Never mind this can of worms. Here I extract the major forces and events that everyone agrees upon based on a synthesis of MIT's 2013 technical review and Blais et al. (2016) (see system diagram, figure 3.1).

Rain and Snow in the Assiniboine Basin: First it rains like crazy in the summer of 2010 all over the Assiniboine Basin. Then in October a ferocious storm of snow and rain attacks all the southern prairies. By freeze-up, the soil is full of water. As winter turns from 2010 to 2011, the snow keeps coming, accumulating higher and higher. That spring brings more snow and more rain. The following summer brings at least four more heavy rain events and some smaller ones. By late April and early May, all the lakes, rivers, and creeks of the basin are at high-water levels or over flood stage. At various locations along the Assiniboine and its tributaries, multiple flood peaks hit all-time records.

In response to the forces of water falling on and flowing across the flood bowl, MIT carries out the following actions (here divided into southern and northern ends of Lake Manitoba).

Operations at the Portage Diversion Structure along the Assiniboine: The spring flood outlook of the Manitoba Hydrologic Forecasting Centre predicts that a major flood event is highly probable. I'm sure all the farmers and ranchers already know this. In addition to all the water saturating the system, everyone is probably worried about ice jams, which can obstruct the large volumes of water expected to flow through the Assiniboine. The Shellmouth Reservoir, which captures Assiniboine waters on the border with Saskatchewan, is drawn down to historic lows over the winter in order to increase storage capacity for spring and help to manage the waters in the Assiniboine before it heads across the provincial border toward Winnipeg, at least for a while. MIT also keeps the level of the reservoir attached to the Portage Diversion low.

On 1 April, MIT opens the gates into the diversion channel to loosen the ice formed over the winter. In early May, forecasters predict that the Assiniboine will peak at levels deemed unsafe for the diversion and the dikes downstream. In response, MIT builds more earthworks to try and increase diversion capacity. On 10 May, fears of dike failure and uncontrolled breach in the Assiniboine channel grow. MIT again increases flows through the diversion into Lake Manitoba. By 11–12 May, the Portage Reservoir is full, and an ice jam on the Assiniboine causes the river to overtop a dike, flooding farms on the river's north side. MIT then opens the diversion to the limits of its design capacity. As May proceeds, in light of forecasts steadily revising upward, MIT steadily increases flows through the diversion, exceeding the limits of its design capacity through 21 May. MIT, in coordination with the Department of Emergency Management, also prepares for a controlled breach at Hoop and Holler (see also chapter 7 for 2014 events). On 31 May, the storm blows in across a very full Lake Manitoba, devastating communities in the southeast basin such as Twin Lake Beach. Nevertheless, MIT persists in diverting water into the lake through 11 July, albeit at a somewhat lower intensity than it had in mid-May. The Waterhen River, Lake Manitoba's natural tributary, continues flowing throughout. Lake Manitoba finally drops below flood stage on 12 February, 2012.

Using the Water Balance Methodology to account for inflow and outflow, MIT ultimately decides in their favour that it would have been worse had no structures been there. The recorded lake level in Manitoba is lower than it would have been in the absence of provincial water control works because the Fairford River water control structure that sends outflows from Lake Manitoba into Lake St Martin more than compensates for the increased height of the lake levels as a result of Portage Diversion operations in May through July 2011. Moreover, MIT calculates that while the lake would have peaked at a lower level, it would have stayed in flood stage "much longer, well into 2013" (MIT 2013b, 112). This assessment happens to relieve MIT of financial responsibility for artificial flooding.

On 31 May, with the lake level at an all-time high, Twin Lakes Beach in the municipality of St Laurent experiences "wind set-up," that is, is hit with strong, sustained winds that forces water into the south basin. "The effect of wind and waves on this date produced the highest ever instantaneous water levels on the south and east shores of Lake Manitoba" (MIT 2013b, 82). But the assessment eliminates wind's peaks and lows, the WEE thereby assigning the disastrous effects of wind and wave to nature.

What would have happened if the Portage Diversion had not dumped a record level of water into Lake Manitoba when levels were already high due to the preceding year of rain and snow? The question that comes to mind is: Why can't things be organized such that the possibility of wind set-up is part of the design equation? Yes, winds are unpredictable in their particular instantiations. But wind set up in the south basin is not completely unpredictable or even unexpected. It can happen, especially in storms. In practice then, the system defines the south basin folks as infrastructural outsiders situated outside the core zone of urban protection. The flood control system demonstrates a technozone's nature in that it "accelerates and intensifies agency in particular directions ... with unpredictable and dynamic effects" (Barry 2006, 241). In this case, one might find that MIT collaborates with wind, accelerating its storm-driven agency.

And now to the far north end, where Lake St Martin outflows into Lake Winnipeg via the Dauphin River: Dauphin simply can't handle all the overflows, a situation complicated by its tendency to form frazil ice. Frazil ice creates ice dams that hang down from the surface into the channel so as to drastically block water flow and reduce Lake St Martin's outflow capacity.

> *Operations at the Fairford Control Structure on Lake St Martin:* By January 2011, Lake Manitoba sends enough water downstream to put Lake St Martin at its maximum desirable level. By spring, Lake St Martin exceeds desirable levels and rises even further with summer flooding. By July, all historical levels are exceeded. Those residents of the First Nation communities around the lake who are willing are evacuated. In late July, the province decides to construct an emergency outlet from Lake St Martin to supplement the natural outflow from the Dauphin River. But by late summer and into fall, the ground is wet and marshy, slowing construction. Heading into winter with the looming risk of frazil ice formation, MIT decides to make a smaller channel than planned and withdraw before freeze-up. It's a mess.

MIT authors start their review of this situation by dissembling. They point out the community's vulnerability going into the storm, i.e., the factors other than their own actions that contribute to extreme events. Lake St Martin First Nation is vulnerable because it sits on low lands around the lake. And too, the Dauphin River has insufficient capacity. These caveats in assessing responsibility relate to, but do not address, concern that such vulnerability is perpetuated by routine management practices between, rather than during, extreme events.[28]

The flood of 2011, although tied directly to the engineering of the Fairford River Control Structure of 1961, may also be seen as the outcome of a string of catastrophes that start with colonization. The Anishinaabe ancestors of the people of Lake St Martin First Nation lived by the lakes and rivers of the boreal forests between Lake Manitoba and Lakes Winnipeg for many generations and have been in situ since the mid-1850s, before the settlers had even formalized the province with the Crown. State-induced precarity solidified in treaties with the Crown gravely reduced territories of all First Nations peoples, including the Anishinaabe people of Manitoba, who were assigned reserves in remote, swampy areas, often in floodplains (Thompson, Ballard, and Martin 2014). And the settlers' sedentary geo-cultural model has made ever more difficult seasonal movements away from rivers in flood and back in time to plant and hold gatherings. In other words, the benefits of Indigenous relationships with water have been diminished by property exclusions imposed by settler law. Loss of autonomy has intensified hazards created by flood control engineering.

Back in the 1890s, when Upham was trolling the basin of Lake Agassiz, the federal Department of Public Works constructed a canal between Lake Manitoba and the Fairford River "for the purpose of improving the outflow from the lake and providing flood relief to settlers around the lake" (MIT 2013b, 53). After the flood of 1950, when Manitoba built the modern flood control technozone that created the conditions for a thriving urban economy near the southern border with the US, hazards to the communities on Lake St Martin intensified in step with the construction of public works on Fairford and then at Portage. The Fairford Control Structure "was never designed to match the capacity of the Portage Diversion at any particular point in time." The design priority was to maintain levels in Lake Manitoba within a desirable range (MIT 2013b, 112). As a result, since the 1960s and 1970s, there have been more frequent and higher flows into Lake St Martin. Climate change is not mentioned.

In effect, and this is my interpretation, the technozone was never meant to protect the people who live in Lake St Martin. MIT has actually been successful if you measure spatial justice according to nineteenth-century provincial priorities. Indigenous legal scholarship and activism provide abundant examples of official government acts in other sectors that strongly suggest that negative consequences in the domain of flood control may not simply be a case of limited effectiveness in the face of implacable geological forces (e.g., see Borrows 2001 and Consumers' Association of Canada 2014). Consider too a late nineteenth-century amendment to the Indian Act "to include a definition: 'person' was defined to mean 'anyone other than an Indian'" (op. cit.). Sheryl Lightfoot (2020, 284), in the conclusion to *Pathways of Reconciliation*, writes that the Indian Act is "the antithesis of self-government as an expression of self-determination." Such legal precedents cast shadows over everything, including flood control engineering.[29]

V. INFRASTRUCTURAL EDGE = INTERNAL FRONTIER: AN URBAN GHETTO IN THE WILDLANDS[30]

The system does not have access to the outside, except through its own construction.
    Andreas Philippopoulos-Mihalopoulos (2015) in *Spatial Justice*

The good people of Winnipeg are all for engineering that keeps land separate from the rivers that run through it, but they do not generally like nor intend to harm other Manitobans by doing so. In today's times, respect for First Nation, Inuit, and Métis peoples, many of whom live in the northern reaches of the province, is a principle that is widely accepted by the populace and enshrined in law (e.g., Canada supports UNDRIP, the United Nations Declaration on the Rights of Indigenous People). And yet, historically and materially sedimented forms of discrimination in everyday life and in law do not dissolve autonomously. Sedimented forms of discrimination act as material givens that confer liberties on provincial agents who extend their logistical power and authority far beyond the urban seat of their offices (Thompson 2015).[31] Embedded in so many layered formations of historical dispossession, it becomes difficult to disentangle or even to be conscious of the way the alliance of engineering and law reinscribes inequality during and after extreme floods. In the wake of extreme flood events, mass media circulate tragic images of displaced humans and homesites; people are not unaware of the problems. However, media tend to communicate engineering's ineffectiveness in protecting all as political failure, as a lack of progress. This frame of interpretation disguises engineering's convenient repetition of a brutal political compromise.

A form of material power inherited from settler-colonialism, engineering's political compromise depends on reproducing Indigenous land within the province as *terra nullius*, or wasteland (cf. Coulthard 2014, 175).[32] In their introduction to *Pathways of Reconciliation*, Craft and Regan (2020a, xii) write: "As part of the ninety-four Calls to Action, the TRC specifically called for the repudiation of the Doctrine of Discovery and *terra nullius*. However, as many have argued, this requires a dismantling of the institutions that have been structured by, and continue to benefit from, these legal assumptions." Flood control is one such institution built on the assumption of Indigenous sacrifice.

In a territory as vast and flood-prone as Manitoba, control, like balance, is an ideal to strive for rather than a feasible expectation. True, individual interviewee-experts share critiques about MIT decisions and actions (or lack thereof), relating how they would have done things differently if ... But my general impression is that, within the everyday realm of their operations and decision-making, most provincial experts mean well and are trying their best. Everyone follows required regulatory standards designed for normal times until extreme circumstances make following normal regulatory standards impossible. Everyone tries to be transparent about the ways they operationalize spatial justice. Their (settlers') sense of fairness and techniques of transparency assume a common legal culture. In this context, the ethnographer's task is to discover how and why certain aspects become public while others become invisibilized and to show how intentional passivity is an active working mode of political engineering. This is what I figured out:

> In the context of what is generally considered to be a successful flood control system, ineffectiveness at the edges generates an internal contradiction. For infrastructural outsiders, the metaphor of the province as trusted administrative authority has broken

down, to the extent that it ever existed at all. Trust can be rebuilt through post-disaster relief. But, as they say, the floodwaters must go somewhere. In the far lacustrine edge of the re-engineered Assiniboine basin, Lake St Martin is the end of the line. It is an internal frontier.

I hypothesize that internal frontiers emerge along three-dimensional geophysical edges that separate the insides (and near-insides) from (far) outsides within a theoretically unified jurisdiction.[33] Building on Philippopoulos-Mihalopoulos's quote above, the only way to create an inside, protected, or semi-protected space is to create an outside, unprotected space. Surveyors draw lines on maps to perform and represent the separate spaces of sovereign entities. But surveyors will not be sent to draw such lines along the most hazardous infrastructural edge for the purposes of performing sovereignty *within* provincial space.

> The internal frontier is a geo-cultural breakdown,
> a cut through the metaphorical real of provincial territory.

Internal frontiers contingent on infrastructure are paradoxical: they are insides treated like outsides, like urban ghettos of the wildlands.[34] Frontiers formed by infrastructural edges may come to seem like arrangements impelled by the threat of water's uncontrollable logistical powers – to this end, Lake Agassiz and the Outburst Quartet might work their mythical powers on the geological imagination.

An internal frontier's material premise creates an atmosphere that disguises the workings of the law that underpin its formation. The atmosphere is a dissimulation in which law withdraws from the landscape (Philippopoulos-Mihalopoulos 2015, 107–8). Far from the halls of legislative wrangling, police or carceral bodies, and architectures, the concrete imperatives of engineering floodwaters pushes the law's presence into the background.[35] And yet it must be the law that allows engineers to fall short, or, at least, the law does not give the engineers enough impetus and resources to imagine and strive to create a spatially just flood control system.

> *In a just world, I argue, experts could wield the law in order to fairly distribute harms resulting from engineering's technical prowess and*

limits. How might MIT's *capabilities in technical measurement and regulation of lake levels be turned toward creating a sense of balance that distributes risk more evenly?*

This is a globally relevant question. Internal frontiers associated with flood control systems proliferate along with the infrastructural models learned and circulated by engineers. The cyber-earthworks are basic elements in the ubiquitous drive to stretch technozones from the cities that depend on them to rural and wildland aquatic spaces that are basically given no choice but to observe and receive, one way or another, the impact of that dependence.[36]

6

Sunlight Machines in Floating Ecologies (Diatoms)

Engineering is always already ecological, even if the ecological knowledge at work is nil.

> Jerome Whitington (2018, 149) in *Anthropogenic Rivers*

There are dimensions of human–water relationships that exist in, beside, or outside the ambit of the flood control technozone, enabling it and affected by it. Widening the circle of elemental consideration to include other kinds of kin and caring, this chapter ventures afield into microscopic life forms and lacustrine architectures (cf. Haraway 2016). After looping back through geological layers of diatoms near the Portage Diversion site, the chapter ends with a spin into the exuberant vertical mix of waterspouts.

1. HOW DIATOMS COME TO MATTER

Matter itself is not a substrate or a medium for the flow of desire. Materiality itself is always already a desiring dynamism, a reiterative reconfiguring, energized and energizing, enlivened and enlivening. I have been particularly interested in how matter comes to matter. How matter makes itself felt.

> Karen Barad (2012, 53) in *New Materialism*[1]

Scientists peer through microscope lenses to learn how lake diatoms come to matter. Diatoms travel through rivers and into lakes, there to enliven lacustrine habitat. Lakes are bodies and homes for other bodies, intertwined spheres of co-evolution and scientific action. When diatom species disperse into lakes, they each seek out niches with qualities that allow them to thrive. For those who have microscopes and seek out lacustrine adventures, the species-specific patterns of diatoms reveal episodes in the geological story upon which the flood control system is premised. When collected, washed, and separated from their slow-floating or well-sunken companions, diatoms lie still on glass slides, indifferent to the currents that bring them into the bodies of knowledge through which humans realize lacustrine ecologies. And yet, diatoms can function as proxies to communicate the effects of flood control, thereby helping humans to realize their lacustrine habitats sustainably.

I approach Barad's problem of how "matter makes itself felt" by tracing the contours of a collaborative space through which scientists collect and convert living and dead diatom bodies into visual images. And I sketch a path through which the taxonomy and ecology of diatoms gets taken up by Manitoban geoscientists who empirically reconstruct the postglacial relationship between Lake Manitoba and the Assiniboine River, the site of the Portage Diversion. This roundabout journey through the study of lakes (limnology) and lake-beings touches down in a transdisciplinary, transhistorical, multi-scalar space that brings geoscience into play with natural history, botany, biology, and paleoecology. Through a series of sketches, I show how diatoms first become visible to science through interweaving technological and theoretical change. I show how diatom bodies make themselves felt as botanical objects and geological artefacts, as buoyant sunshine machines and dead particles in geological cores dug from the flowing waters and dark underbellies of lakes.

Unlike erratics, the giant rocks that travel with glaciers, diatoms are not prominent geo-cultural actors. Most people probably think of them, to the extent that they are thought of at all, as sensitive botanic materializations of an inert world waiting to be discovered and put to good use. The intricate patterns of their species-specific bodies become visible through microscopes they have helped to invent. And nuanced patterns of their distribution in lake space quietly fortify the scientific rationale of the Portage Diversion – an engineered connection with natural precedents.[2] I discuss their usefulness while aspiring all the while to convey the "enlivened and enlivening" lightness of diatom-being. Except for scientists without whom diatoms would not have a story to tell, I decentre most humans so as to honour the magical, beautifully indifferent qualities of diatom life.[3]

II. LAKES AS EQUILIBRATORS OF CHANGE IN RIVER NETWORKS

The core linear impulse of rivers moves beings and things from one place to another. But once water flows from river to marsh and into a lake, water tends to assume the form of a destination, a welcoming habitat-in-flux within which beings and things may find it auspicious to dwell for a time. Though rivers both fill and empty Manitoba's largest lakes, essentially flowing through them, geo-culturally, lakes also act as autonomous creature-places.[4] Toes wriggling in sand, inhabitants experience lakes as historically changing views, routes, depths of immersion in and across open blue-grey space. In their work, scientists experience lakes as objects in and from which knowledge can be inferred. The earthworks of engineers are grounded in the reliable details and uncertainties of this basic geoscientific knowledge. And geoscientists continue to learn a great deal about Lakes Manitoba and Winnipeg, in part because of their significance as water storage infrastructures in flood control and hydropower, respectively.

Long after fieldwork, when I get ready to write about lakes as a specific form of water body, I look up the definition in a physical geography text that hints at the difficulty of interpreting elemental flows (Strahler and Strahler 2005, 479, paraphrased here):[5]

> Lakes are earth-bound bodies that store volumes of water. Their volumes fill the spaces in bowl-shaped depressions, their surfaces unite earth and sky. Along lacustrine-atmospheric interfaces, *lake levels rise and fall in different timescales simultaneously*: in seasons; in periods of heavy rain or drought; in multi-decadal cycles of wet and dry; in climates changing over geological time. Within any given timescale, the water balances of lakes depend on how much they receive from precipitation and inflowing rivers and streams and how much they release through outflow and evaporation. [my italics]

Then I find an article about rivers as "integrators of change in the hydrological cycle," co-authored by J.P.M. Syvitski, a geoscientist with a passion for situating geoscience in transdisciplinary theories and methods (Overeem and Syvitski 2010, 285). I write my friend and colleague Syvitski an email asking if lakes might also be considered integrators of change in the hydrological cycle (and by extension in the flood control system). The response:[6]

> Good question. Lakes are also integrators, like all parts of a river network. I would say that they are also dampeners and modifiers. If a flood wave sloshes down a river, the wave will flatten over a lake and dampen. Lakes are also modifiers because the residence time of a water particle is greatly enhanced, allowing sediment to settle out, diatoms to flourish, carbon to be sequestered, and a lake ecosystem to equilibrate and exchange with the river ecosystem.
> .... equalizing bodies?
> .... equilibrators of changes ...?

Opening up the lake's body to the hydrological dynamics (and by implication the flood control system), Syvitski's reply crosses boundaries between living and dead matter and spacetime scales. Fleshing this out, one could say, then, that a lake's integrative power can be characterized by its unique potential to form a body with a level surface that, when healthy, induces sufficient calmness to create nutrient-rich ecologies of light. In slowing down the waves sloshing down a river, lakes give time to sediments to settle, light to penetrate, silica to organize in exquisite one-celled formations, diatoms to thrive. Creating ecological habitat for living things capable of transforming light into life and feeding the world, including, of course, all the networks of scientists in their many disciplinary niches.

Besides inspiring a geo-savvy lacustrine vernacular, my exchange with Syvitski illustrates how conversation can cross disciplinary boundaries; like crossing cultures, ideas develop as they circulate. This can lead the ethnographer to pivot her attention into unexpected channels. In this case, Syvitski's passing mention of diatoms triggers my curiosity. It sends me into scientific texts on the microscopic worlds within lakes. This side-trip into the "unthought known" of engineering (Bollas 2008, 19), though far from the factors that guide engineers' operations when they divert river floodwaters into lakes, is nevertheless directly within the sphere of their impact. When the control system sends quantities of water into lakes, raising their levels, they also alter their qualities in the short term (e.g., acidity, salinity, turbidity, temperature, nutrient levels, etc.). In the long term, flood control systems alter the general character of the habitat and composition of species. In a sustainable model of governance, lakes would function as storage tanks for flood control systems only if their vitality as habitat for one-celled and collective creatures like diatoms is explicitly, even if indirectly, protected. So consideration here is warranted. And yet ...

Admittedly, diatoms are unusual ethnographic actors, not even the stuff of myths and legends. They do invoke a textual world where more-than-humans can continue to live apart, in our imaginations at least. I do hope, though, that the planetary surface will always have some diatom-rich lakes that never see a human shadow fall across their surface. Slipping off the edge into the water, I am attracted to the diatomaceous pull of unacknowledged belonging.

III. SUNLIGHT'S MEDIATORS AND SCIENTIFIC PARADIGMS

Diatoms are inherently suited for vouchering ... [for] future generations interested in reexamination in light of new discoveries.
Julius and Theriot (2010, 17) in *The Diatoms*

Diatoms thrive in the sunlit equilibria of shallow lakes on calm days between storms.[7] Freshwater diatoms situate themselves in evolution and in daily life by using acute sensitivities to the affordances of varying lake levels. Designed for buoyancy and light infiltration, diatoms float just under the surface with other phytoplankton where sunlight readily penetrates. As diatoms rise and fall with lake levels, they transform sunlight into high-energy lipids by fixing carbon. Two opalescent glass walls, one slightly smaller than the other, are tied with a girdle to form finespun boxes (called frustules) to hold their chlorophyll-loving genes. They float as individuals but may form colonies. Diatoms become food for animals as tiny as zooplankton and in the marine environment as enormous as whales. Diatoms sink when they die, collecting in layers over time to be discovered by the likes of paleoecologists.

Like birds, diatoms have survived some 200 million years longer than dinosaurs – long enough for humans to incorporate their fatty acids when they eat fish and shellfish. And simultaneously, they are credited with absorbing as much carbon dioxide, a major greenhouse gas, as all of the world's tropical rain forests.[8] In this time, researchers have only studied a small fraction of their unique evolutionary lineage, which has diversified to a degree rivalled only by insects. But their lineage is not as ancient as their great diversity might suggest. In fact, "the origin of this first diatom is approximately 60 or 70 million years younger than the specialized teeth found in mammals, including in the reader's mouth" (Julius and Theriot 2010, 18).[9]

Diatoms bloom (and hang) in spring and burst (and sink) in rhythmic response to changing conditions. Dead and alive, in fresh and salt, motile and adhering to substrates, planktonic and benthic, diatoms reveal themselves as collectivities and individuals. Pollution from sewage and agricultural runoff can turn them into gunk and shift the balance of lacustrine ecology. To those who can see these organisms in taxonomic and ecological relationship, their patterned presence reveals not only the power of lakes to slow waters moving through the hydrological cycle, as Syvitski indicates, but, too, can tell the forces of change that negatively affect lakes. Indeed, the study of diatoms and their application is a scientific enclave replete with beautiful and precise images of the diverse and elegant shapes of its more than 200,000 species distributed throughout the world's waters.

Diatoms are primary producers of the lacustrine food web; they also feed scientific knowledge webs. The orderly images, descriptions, and taxonomies upon which scientists dwell betray a fascination with these primary beings whose liveliness lies beyond the range of unaided vision. To capture them, diatom studies develops a unique relationship to technologies of light and vision and, too, discovers novel ways to combine discovery and invention, the twin forms of divided scientific knowledge (cf. Strathern 2005, 39). I cannot help but think that the prodigious collection of life form images might be driven as much by the power of aesthetic attraction as by taxonomic or biological significance.[10] And I wonder how we humans might acknowledge the agency of diatoms as unintentional collaborators in our discoveries and inventions.

An archive-influenced riff on a scene of intermingled agencies gleaned from scientific texts:

> *In seventeenth-century Netherlands, Leeuwenhoek hand-grinds single lenses in his shop:* When he places the lenses in just the right light, he can magnify one-celled organisms up to 200 times. He can see microscopic creatures! He calls them Animalcules. Leeuwenhoek's eyesight is keen, but he is not adept enough at drawing to make it possible for later generations to be sure whether his Animalcules

represent the first recorded diatoms (cf. Round, Crawford, and Mann 1990, 1–2).

When the powers of microscopes grow enough, diatoms reveal the chlorophyll they hold inside. So it is decided – they are not animals but, rather, plants (Round et al. 1990). But long after botanists become secure that diatoms belong in their dominion, their classification is queered once again. When their nuclear genome becomes visible, it turns out that diatoms, like animals (the metazoans), rely on a urea cycle to manage nitrogen flux (Allen et al. 2011; European Commission 2011).[11] Evolution is a monster.

IV. GEOLOGICAL BOUNDARY-CROSSINGS: LAKE MANITOBA'S DEEP TIME DIATOMS

These *ecosystems*, as we may call them, are of the most various kinds and sizes. They form one category of the multitudinous physical systems of the universe, which range from the universe as a whole down to the atom.
 A.G. Tansley (1935, cited in Lindeman 1942), both in *Ecology*

Tracing the original ecosystem concept back to the early nineteenth-century study of lakes, limnology, Lindeman (1942, 400) defines ecology as "composed of physical-chemical-biological processes active within a space-time unit of any magnitude," i.e., the biotic community *plus* its abiotic environment. I find Lindeman in Bennett (2004, 349, 353), whom she quotes as part of her quest to understand the "ecology of matter," the "affinities between organic and inorganic matter."[12] Separated by discipline and more than six decades of scholarship, their work converges at the point where it becomes possible to integrate different domains of liveliness. I build on this holistic, transdisciplinary sense of vibrant matter to write about what empirical mysteries diatom-rich shallow lakes reveal. While the fatty acids that living and breathing diatoms make from sunlight nourish their fellow water-loving creatures, it is the dead diatoms that excite geologists and paleoecologists.

Research networks build up around diatoms from Manitoban lakes, the shallowness of which makes them a great place to study their taxonomy and ecology.[13] Diatoms also make Lake Manitoba a great place to study postglacial geoscience. And Lake Manitoba geoscience is a great place to search for precedents and rationales for flood control policy and law. That said, this list of three overlapping research networks produces what Tsing would call "incompatible data sets," out of which emerges a legitimate critique.[14] University of Winnipeg freshwater biologist Eva Pip argues that government reliance on the lakes as storage basins for flood control and hydroelectricity, at the same time that their more-than-human inhabitants attempt to cope with the imposition of phosphorus and nitrogen-rich sewage, directly contributes to the fragility of lacustrine ecosystems, "pushing them beyond the point of no return."[15]

Diatom architecture offers qualities of complexity and longevity to observers. The delicate variations of diatom frustules, their species-specific silica casings, can last millennia, long after these beings participate in living ecologies. In other words, the uniqueness of frustule architecture has classificatory powers that index deep time ecological and geological relationships. Freed from the need to study living diatoms, which can only be seen with light microscopes, scientists can rely on the much greater visualization powers of electronic microscopes. Once organic matter is removed, microscopists can recognize finer and finer morphological distinctions in the inorganic matter and thereby create more elaborate taxonomies at the species level.

Diatom fossils are doorways into deep time. For much of the Holocene, an arid climate dominates the terrain abandoned by Lake Agassiz, but in the epoch's last few millennia, moisture increases again. Even in some of the arid times, however, Lake Manitoba queerly fills with fresh waters. As it turns out, the fresh water comes not from rain or snow but from the Assiniboine River. Geoscientists know this because of diatoms. During all these millennia, diatoms have been moving and adapting to the changing shape and salinity of their lake habitats (Bennion et al. 2010, 152f; Risberg et al. 1999, 1,300). When they die, species identity + location register the lake's salinity pattern at that moment in time. By tracking discontinuities among species assemblages, twenty-first-century geoscientists can infer changing configurations of deep time terrain.

The saltier the lake, the more diatom ecology is dominated by taxa with broad tolerance for saltiness. It would take large changes in salinity to change the species composition of the diatoms that thrive there. They're tough. In contrast, at freshwater/saline boundaries, where the lake tends to be brackish, only some diatom species live comfortably, feeling at home in a narrower, not-too-salty range (like *C. quillensis* and *C. agassizensis*; see below). Geoscientists are more likely to find shifts in species composition at such freshwater/saline boundaries. This kind of diatom data delineates what happens to the terrain that emerges after the Outburst Quartet.

Another archive-influenced riff on a scene that might have happened something like this:

<u>1993, the 12th North American Diatom Meeting, Delta Marsh Field Station, Manitoba</u>: Hannelore Håkansson, the keynote speaker from Sweden, stops in front of a poster bearing disconcerting news. The poster, by Jan Risberg and Hedy Kling, reveals taxonomic confusion among species in the genus *Cyclotella* (first described by Kürtz in 1844) that were collected from slightly brackish fresh water and inland saline water. Håkansson has contributed to this area of inconsistent scholarship in the past. A conversation ensues.[16]

Inspired to collaborate, the two women, Håkansson and Kling, embark on an examination of new core material from the 1963 Lake Manitoba collection and a reexamination of some typed material that might have been misidentified. When they look at the *living* material through a light microscope, they find that *C. agassizensis* can be easily mistaken for *C. quillensis*, a diatom species quite important in Lake Manitoba prehistory (see below). But when they look at three-dimensional images of *dead* material through the scanning electron microscope (SEM),[17] they discover fine differences in frustule architecture. This establishes a significant distinction between the two species.[18] They quickly publish their findings and accompanying micrographs in *Diatom Research*, where I find it more than two decades later (Håkansson and Kling 1994).

The frustule differences that emerged in the comparison of SEM images provoke a classificatory elaboration of the *Cyclotella* genus – one minor discovery (speciation) predicated on a major invention (SEM). Returning to Barad's feminist-physics-philosophy question about how matter comes to matter, I wonder about the relationship between the source medium's ambiguity – brackish water is not quite fresh, not quite salty – and the inconsistency of taxonomy. Like Håkansson and Kling, my instinct is to take inconsistency seriously. However, rather than always sorting things out, I also want to be able to freely appreciate diatomaceous queerness, i.e., the ability to move among or hang out between categories.

A. *Salinity and Speciation* (Cyclotella)

Once their silica skeletons sink, the intricate forms of diatom fossils endure in the sedimentary layers of deep timespace (Risberg et al. 1999, 1,303–4, 1,312; Last 1984). The mineral magnetic resonance patterns coding their past existence provides geologists with the power to pin down the changing character of Lake Agassiz's post-glacial transition into Lake Manitoba. Following a set of complicated methods,[19] Risberg et al. identify 144 diatom species and group them by ecological predilection (thirty-two are halophilous, twenty-eight are brackish, thirty-three are indifferent, twenty-three are freshwater, two are aerophilous, and twenty-six are unknown). Among the *Cyclotella*, the list of brackish diatoms includes *C. agassizensis*, with attribution to Håkansson and Kling, the first list to distinguish the species as new. Together, the species groups form marker zones and assemblages in geologists' lithography columns and micrographs. The collaboration between diatoms and geoscientists produces knowledge on the shifting relationship of the Assiniboine River and Lake Manitoba:

> Between 6,000 and 4,000 BP, the Assiniboine River drained through one of its paleochannels, infusing freshwater into a somewhat brackish Lake Manitoba, and then, between 4,000 and 2,600 BP veered off through another paleochannel, leaving Lake Manitoba to resume its hospitality to brackish, rather than fresh, diatom species. By the late Holocene, the climate got cooler and wetter even as the glaciers farther north continued to melt, causing the terrain closer to the Arctic to tilt up and, in tandem, the terrain closer to the south end of Lake Manitoba to tilt downward in a process called isostatic rebound. These climatic and geophysical process led Lake Manitoba to freshen up and to resemble its current state.

Now we know.

V. MIXING IT UP WITH WIND

In the material semiotics of lake life and light, scientists and the creatures they study create bounded "patches of order in a sea of disorder." Some of the most interesting places in these patches of order lie "where different orders rub up against one another" (Law 2009, 144, drawing on Serres 1974).

Rivers in flood sweep their muddy waters over the marshes and into the lake, working with wind and wave. The riverine order may banish the quiet stillness and clarity conducive to the ordered lake assemblages of diatom species. But as it turns out, this can be a good thing. Floating ecology works rather like this:

A. Winds Counteract Stratification in the Water Column
Diverse and productive diatom assemblages fuel unique food-web pathways in lacustrine littoral zones – but only if winds can help them stay buoyant enough to catch the sunlight (Bennion et al. 2010, 153). Shallow lakes are often subject to intermittent mixing events that resuspend diatoms, lifting them into the photic zone and promoting population growth. The relatively dense silica of their cell walls tends to make them heavier than other multi-species plankton assemblages. Fish swimming around also help diatoms to stay afloat. In short, diatoms thrive with disturbance, not equilibrium. But just in case the winds and the fish are not active enough and the diatom population shrinks to a dangerously low level, they do have the option of switching from their usual asexual mode of reproduction. To bring the numbers in their assemblages back to healthy levels, they start having sex again!

Diatoms always seem to have back-up plans. Not only do they have plant-animal, sexual-asexual capabilities, but several species can exist as meroplankton, that is, they have the ability to form resting cells that withstand periods of prolonged darkness. They can reside on lake bottom and then get reintroduced into the water column during high-wind periods, thereby reseeding the plankton. This capability has a marked effect on phytoplankton biomass and on taxonomic composition and might explain a significant portion of temporal variation in shallow, productive lakes like Lake Manitoba.[20]

B. Floods Alter the Qualities and Connections among Water Bodies
Diatoms can travel. In flood events, distinctive diatom assemblages may converge and sink into the scoured floodplain. Those that like deep water may wind up in the shallows and vice versa. Thus, species composition and location of sediment assemblages register flood frequency and severity. Diatoms can thus function as indica-

tors of long-term environmental change, their species-specific sensitivities enabling scientists to track different processes. Like giant erratics picked up in the formidable grip of glaciers that then disappear in retreat (see chapter 2), diatoms move with the energy of wind that comes and goes. In their respective temperature-inflected milieus, where erratics and diatoms come to rest – their precise out-of-placeness – is the source of science stories.

In the lacustrine world of sunshine machines, winds and floods bring change, at times gently, at times chaotically, and in some places, change is brought about when water is rerouted artificially by large infrastructures like the dam and spillway at the Portage Diversion (see chapters 1 and 3). But in the long run, diatoms have thrived in the mix.

VI. THE SUPERNATURAL EMPIRICS OF WATERSPOUTS: VERTICAL EXCESS

There is a subtle entanglement and confusion between all beings on earth, a consequence not only of our ancestry, and the cellular similarities of our makeup, but also of our subjection to variant aspects of the same whirling world.
David Abram (2010, 192) in *Becoming Animal: An Earthly Cosmology*

The level-ness of lakes is fundamental to their existence as bodies and flows and the sense of everydayness they evoke. Yet their wet horizontality can provoke awesome verticalities when interacting with atmospheric agents. When wind reorganizes flowing lacustrine shapes, its energetic patterns may induce confidence in their reliability (to sailboats, for example) or fear of disaster (sinking them).

Look at Lake Winnipeg, Lake Manitoba's big sister, for a dose of the supernatural:

The bodily potential of a large lake can act like a sea, as we know from flood-provoking storms. Their violent, capricious, and practically unchartable character when interacting with the wind manifests most dramatically as waterspouts. Less destructive than tornadoes, waterspouts are practically dustless and frictionless, their winds driven more by humidity than heat. "Byways of wind," Guy Murchie calls them in *Song of the Sky: An Exploration of the Ocean of Air* (1954, 97).[21]

A pilot and writer, Murchie combines science and poetry in his telling. He describes waterspouts out at sea (155–6):

> They often turn out when the sky is sultry and sullen, when there may be a bluish haze and surface suggestions of the moodier air currents. Suddenly a dark elephantine cloud with a hanging trunk is noticed, usually lengthening rapidly from the stubby tornadic funnel to the scrawnier, snakelike stem of the typical waterspout. As the winds carry it along, the thin whirling stalk may bow forward ahead of its cloud dragging its spinning base after it willy-nilly, like a celestial vine goddess borne by her torso.

Where there are large lakes, atmospheric encounters can pull the waters right up from their beds. Manitoba lies in the northern temperate zone within the ambit of prevailing westerlies, a broad sash of cool, dry winds that extend down past the forty-ninth parallel into the US and across the Atlantic Ocean to Europe and Asia. When the westerlies encounter the hot, moist air sweeping up across the Great Plains from the Gulf of Mexico, Lake Manitoba and Lake Winnipeg may host a lot of storms, some with flocks of waterspouts. Manitoban summers are hot, and the large shallow lakes produce updrafts of hot, humid air.

In her book about Lake Winnipeg, *Mistehay Sakahegan: The Great Lake*, Frances Russell (2004, 28–9), presents several witness accounts. They are accompanied by photos that are consistent with Murchie's narrative description above. Here is one:

> Moving across Lake Winnipeg on 8 August 1984, "At least six and perhaps eight 600-metre tall waterspouts made a stately 25-kilometre procession ... Some were thin, black threads while others were thick, gray or gleaming white pillars. They dangled from a flat-topped, ragged bottomed black cloud and marched soundlessly across the water. Around the foot of each was a circular fountain composed of a wall of water and spray about a metre high. Two of the spouts embraced each other as they traveled ... What witnesses mostly recalled at the time was the utter silence."

Thus, thunderstorms and strong winds can force lakes to abandon their flatness, triggering dramatic transformations that transfix observers. Although photographers may capture them, they elude the systematic measures of scientists seeking to actively observe. Waterspouts mark the outer edge of lacustrine reliability. But imagine, too, the diatom assemblages that meet and mix in the embracing spouts as they travel, only to fall with surprisingly random grace into another floating ecology.

7

River-Trickster on TV
(2014 Flood)

Surging rivers have the power to arrange language as well as landscape. They give character to the geological imagination not only as familiar elemental figures, like mermaids or goddesses, but rather as material transfigurations of spacetime itself – tricksters of the Anthropocene. I analyze water's elemental powers in governmental and literary forms, moving from a televised press conference convened during the Flood of 2014 to a novel by William Faulkner and plays by Ian Ross and Tennessee Williams. These flood genres have performance contexts and specific social functions that I analyze ethnographically as geo-cultural evidence of shared states-of-being. But their ultimate affects escape the ethnographer's grasp. Boundaries dissolve as they travel in the sphere of unintended agencies, where the rivers' elemental powers inspire a bubbling bricolage of geoscience and flood fiction, politics, and poetry. Blended habitus merges linguistic forms and flood moments and, I suggest, carries them into the *longue durée* where they linger as the unthought known. There is a great deal to learn, however, from the expressive language of flooding landscapes before they escape the means of empirical capture.

I. LOGISTICS IN MOTION: SUMMER FLOODING ON THE ASSINIBOINE

A. *As a Flood Approaches, Queerly*

In days of emergency approaching the worst of the July 2014 flood, the media landscape and the geophysical landscape converge, revealing public performance as an essential geo-cultural responsibility of engineering officials. Once a flood crisis is underway, top engineers, politicians, and emergency experts shift into coordinated public performances that translate technical operations into the vernacular. To inform the public and build consensus about their intention to enact particular forms of logistical power, officials manage emotional suspense by carefully crafting demeanour and discourse. In the press conference I analyze in this chapter, they announce a State of Emergency that triggers military intervention. They focus on their immediate preparations for a contentious dike breach maneuver on the Assiniboine River near the small (settler) farming community of Hoop and Holler.

Three years earlier, in the flood of 2011, the Assiniboine River first announced the possibility that climate change had already begun to alter regional flood patterns. Historically, and at least since the era of modern infrastructure, the river had always played a supporting role to the much grander and fearsome Red. Its dikes were never built or maintained to withstand the 2011 extreme flood – "the largest flood event on the Assiniboine River in living memory" (see chapter 5). And troublingly, when the Assiniboine came toward Winnipeg in May 2011, it did so in synchrony with a flooding Red, multiplying the hazard (MIT 2013b, 38).[1] It did not let up when it reached Manitoba's border with Saskatchewan, where the Shellmouth Dam and Reservoir regulates Assiniboine waters. From there it meandered across the prairies until the Portage Diversion, where engineers diverted a contentious portion of its flow before allowing the remainder to continue downstream to Winnipeg. Also, troublingly, the flooding was prolonged. The snow-driven flood went on so long that it became a rainwater-driven flood.[2] Afterwards, MIT did a technical review and stepped up dike maintenance.

This chapter picks up flood history three years later, in 2014. The Assiniboine and its tributaries rose with extraordinary force, not just again but queerly. This time, the flood started in the summer, not in the spring as usual. MIT's engineers and provincial politicos were nearly caught off-guard. They had the skills, resources, and a plan of action to address these familiar-yet-odd circumstances, of course. And they committed themselves to actively demonstrating these logistical powers, materially and symbolically.[3] Hence the press conference. Staged in the initial hours of the 2014 flood, the press conference announces a state of emergency triggered by a worsening forecast as a result of rains from the west.

To analyze the press conference as an event performed in a particular genre, I borrow the concept of *chronotope* (= timespace of narrative). Bakhtin's (1981, 84–5) chronotope refers to the artistic process of assimilating and appropriating real historical timespace in literary genres.[4] I apply the concept to the press conference, which although not artistic in the usual sense is artfully crafted. In the context of extreme uncertainty, officials carefully craft the information at hand to create an authoritative narrative representation of emergency timespace. They aim the provincial intention to control at the public's anxieties and fears as much as at geophysical flows.

I show how officials deftly balance uncertainty and deliberative action at interfaces between inhabitants, river, provincial government, and military emergency responders. I find that officials accomplish a sense of balance in their discourse by dividing up their collective communique into distinct *temporal* jurisdictions. These jurisdictions differentiate time-specific forms of expertise, that is, skills in interpreting river past, decision-making in the present, maneuvers planned for the near future. Between the timespaces of the press conference itself (a political event) and the timespaces of the Assiniboine flood outside (a geophysical event), the speakers articulate a set of ordered meanings pertaining to uncertain river-city states-of-being.

Water engineering inspires a method of dividing text into packets for the purpose of analysis. I mimic the measure of cubic feet per second (cfs), the small abstract units of water volume used to estimate and reroute river flow in flood forecasts and operations. Like the flow of water broken down into volumetric units, I break up the flow of discourse into textual units, which allows me to carry out a structural analysis that reveals underlying logic. My goal is to empirically capture the negotiation of uncertainty among engineers who are witnessing and responding to the hydrological surprises of rain and river. The press conference, I suggest, provides a window into governance as an aspect of geo-culture and local TV as a crucial medium for negotiating the space of political engineering.

Changing Patterns: The nature of riverine flooding is recursive. Rivers wax and wane, thaw and freeze according to seasonal variations of temperature and wind. The logistical power of flood control mimics nature's recursivity. System design is predicated on blended habitus keyed to a seasonal cycle that peaks in spring. Blended habitus is also keyed to the directionality of river flooding, moving along south-to-north and west-to-east axes. Historically, Winnipeg's worst floods have come in the spring from the south on the Red, the meltwaters of the Dakotas rushing toward the still-frozen riverbeds of Manitoba. Hence the Floodway, which moves the Red's floodwaters around the city, has been the city's most significant infrastructural investment (see chapter 3). In short, patterns of seasonal recursivity and directionality have provided an envelope of predictability certain enough to warrant huge investments, especially when combined with the impetus to protect increasingly valuable urban property. But by 2014, the envelope of predictability with which forecasters and engineers have become accustomed no longer seems to contain the agency of the west-to-east flowing Assiniboine.

Winnipeg's system is designed to predict and control spring, not summer, floods. The deep winter cold provides stability and relief from the "who knows what will happen" during and after ice break-up in the spring. And in the heat of the summer and in the fall, Winnipeg engineers can, if not hibernate, more or less relax into routine preparations. So sudden summer rainstorms interacting with the Assiniboine and sending floodwaters down through Saskatchewan and into Manitoban prairies can be disconcerting and dangerous. Summer floods make it more difficult to predict and reroute fast-paced river action. In addition to the flood control system's design and operations that may increasingly be out of synch, climate change has upended complacency about basic weather prediction. Historical patterns of stable temperature and precipitation ranges (stationarity) have provided forecasters with a reliable platform against which to measure and extrapolate changes as they unfold. But stationarity is eroding, and there is no clear path to calculating place-based versions of the "new normal" (Milly et al. 2008; Szeto et al. 2015).[5]

Globally and on the North American prairies, extreme floods combine to signal the need for mitigation of and adaptation to what has clearly been recognized as a climate crisis triggered by human activity.[6] Although not explicitly mentioned by officials in the press conference, some notice of possible climate change links may be inferred from references to the peculiarities of the severe summer flood on the Assiniboine in 2014 (see analysis below). Nevertheless, forecasters and engineers are stuck interpreting, regulating, and politically managing the Assiniboine within the limitation of the present control system and its supplements. They work with what they have.

II. STAGING A CONTROLLED BREACH AT HOOP AND HOLLER

The TV medium provides officials with a public stage that enables political engineering in geophysical extremes. The genre of the press conference in particular allows speakers to condense and control the representation of specific emergency actions in an authoritative style. Their live performance is a timely and informative intervention with the aims of demonstrating transparency and accountability and of building at least a tacit consensus among the majority.[7]

What everyone around here knows: If and when the Assiniboine reveals that the scale of its peak disorderliness will allow it to burst out beyond its engineered confines, excess water will rush across the flat prairie into homes, farms, businesses, roads, culverts, and electrical, water, and sewage infrastructures, hitting some places hard and others not at all. So when then-Premier Greg Selinger wraps up his formal introduction with: "Our priority remains insuring the safety of Manitobans and doing everything we can to protect homes and businesses [1:00]," everyone understands that he is glossing over the fact that the safety of Winnipeg is the priority and the proposed sacrificial zones will be managed as fairly as possible.[8] What makes this press conference interesting is the dance between a trickster-river and a pragmatic provincial government, both actors who may or may not put some Manitobans in the way of the flood.

The Assiniboine's resistance to engineering, its threat to elude appropriation and control, calls for and legitimates a scale-up in the application of governmental power. A State of Emergency had been declared in the morning hours before the press conference. Earth-moving machines and soldiers have begun to mobilize. In this liminal moment between prediction and event, the forecast itself – a continually updated techno-cultural artifact subject to falsification (Derrida and Stiegler 2002, 151) – does a surprising amount of rhetorical work.[9] I came to appreciate the possibilities of this rhetorical work in the course of transcribing and analyzing the videotape.

What, beyond defence, are the geo-cultural implications of flood control? Flood control magnifies the salience of the regional geological timespaces within which rivers act. As rivers galvanize public attention, the press conference punctuates the unfolding crisis as the river city's entwined embodied form(s) compulsively swell and withdraw.[10] Logistical power is on display. The immanent likelihood of the Assiniboine wildly blowing past its dikes helps to rationalize investment in engineering, education, and bureaucratic resources. It also bolsters widespread, if sad, acceptance of directing harm toward the infrastructural outsiders in rural Hoop and Holler. To the degree they can show that they can be successful in protecting the urban majority and compensating those intentionally put in harm's way, MIT and the provincial government will benefit.[11] And too, on a practical note, people are familiar with the press conference as the way to get official information in an emergency.[12]

A. Press Conference as Chronotope

The set-up, 4 July 2014:[13] As if torn from their hectic offices for a quick but essential interchange with the public, four poker-faced white men wearing neat shirts or tees unbuttoned at the collar, their white hair or baldness testifying to deep professional experience, sit in a row along a narrow table on an elevated stage. Below them, a live audience not identified or visible to video viewers listens quietly. The table holds mikes and glasses of water, which the panelists drink from on occasion. Some read from pieces of paper. Behind them, large provincial and national flags set off the grey curtain background. The iconic bison on the Manitoban flag directly behind the premier provides the only visual referent to more-than-human nature visible within the frame. A repeated element of provincial iconography, the bison situates this setting within a diverse array of Manitoban enterprises and institutions. On-screen visuals such as inundation maps, although discussed, are notably absent from the space framed by the camera. Their collective speech act initiates the State of Emergency in the public domain and ends sometime in the near future when the river's two expected crests would pass through Winnipeg.

Premier Greg Selinger opens the conference by reading from a prepared script in measured tones that allow him to look up with each sentence or so:

> Good evening. I'm joined by the Assistant Deputy Minister for Water Control and Infrastructure, Doug MacMahon; his Executive Director for Hydrological Forecasting, Steve Topping; and the Assistant Deputy Minister responsible for Emergency Measures, Lee Spencer.

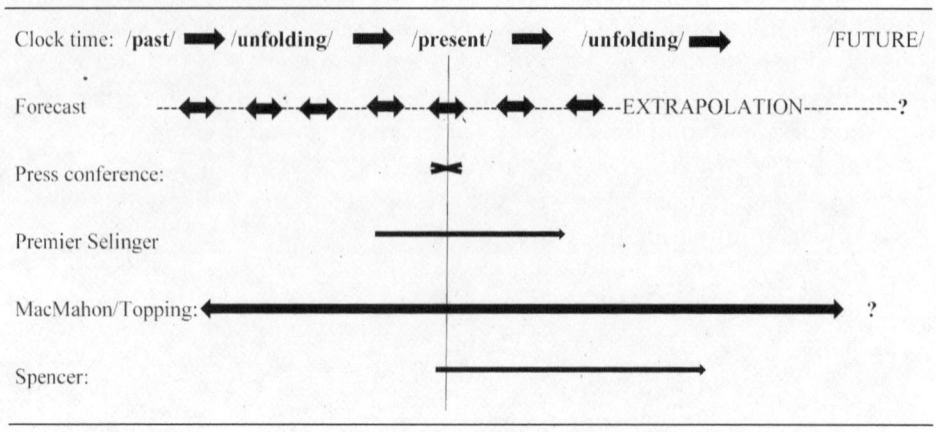

Table 7.1
Temporal jurisdictions within clock time

Clock time:	/past/ ➡ /unfolding/ ➡ /present/ ➡ /unfolding/ ➡	/FUTURE/
Forecast	⬌⬌⬌⬌⬌⬌⬌ EXTRAPOLATION	?
Press conference:	⋈	
Premier Selinger	⸺⸺⸺⸺⸺⟶	
MacMahon/Topping:	⟵⸺⸺⸺⸺⸺⸺⸺⸺⸺⟶	?
Spencer:	⸺⸺⸺⸺⸺⟶	

Temporal jurisdictions: The seating arrangement encodes the provincial government organization chart: Premier, the only one wearing a white business-style shirt, in centre position as political head, with logistics actors arrayed around him. To his right, in internal hierarchical order, sit MIT engineers; to his left, the ex-military man now civilian head of Emergency Measures Organization (EMO).[14] In my analysis of the official transcript, I find that speakers take responsibility for a different *temporal* jurisdiction, i.e., the time frame within which each enacts their responsibilities (although they probably would not think about it in these terms).[15] (See table 7.1.)

The performance dramatizes a logistical pivot point between the morning's updated operational forecast (hydrology = river flows in the immediate past) and the disaster mitigation plan (hydraulics = plan for flood routing). The forecasts, technological artifacts ruled by clock time and composed of geophysical data spanning from 2011 to the near-future-unknown, represent the real. The engineers, who compose the chronotope core, insert forecasts into their river communiques. The whole breaks down into a triple communique, each with its own function and emphasis: explicit political engineering by the premier; forecasting and system operational logistics by the two MIT engineers; and military logistics by the emergency manager. After doing the introductions, the premier first articulates a succinct layperson's version of the now "worsened" forecast. The forecast has allowed MIT to focus on the timespace of the river's possible actions and to propose defensive measures. By including only established facts in the form of the updated forecast and action plan, the premier manages to entirely exclude uncertainty from his part of the communique. He summarizes the logistical aspects most salient to those I identify as "infrastructural outsiders" who may be, in a "controlled" way, intentionally put in danger by the province or who have already unintentionally been put in danger by lack of maintenance along vulnerable dike sections:

> Preparations are underway for a controlled breach at the Hoop and Holler bend to reduce the high possibility of an uncontrolled breach. We have updated the Canadian Forces with these new projections and the need for more resources, more quickly. In the last hour, I have spoken to officials in the affected municipalities.

B. The Plans:
Machinery, Military, and Sandbagging Volunteers

The action plan relies on provincial operators of earth-moving machinery and on federal soldiers, property-owners, and other volunteers for filling and moving sandbags. This becomes clear in the statements of the two MIT engineers and emergency manager. The plan includes two kinds of preventive earthworks: 1) readying a controlled breach in the dikes along an oxbow of the Assiniboine River east of Brandon in order to forestall an "uncontrolled" breach by the river – a defined task with fairly predictable outcome in contrast to 2) policing more than 100 miles of dikes between Brandon and Winnipeg and identifying and shoring up holes (or "hotspots") before the river finds them.[16] The flooded Assiniboine can undermine its dikes, causing an uncontrolled breach by flowing over them and eroding the soil below (overtopping), a problem that can be prevented by building the dikes higher. The river can also cause an uncontrolled breach by bubbling into the soil from below flowing out the dry side (mud boil), a problem that is difficult to detect, or by keeping the channel full for so long that it causes soil saturation (washout), which everyone can watch happening but is difficult to control. The race to find vulnerabilities in dikes is a rather diffuse and labour-intensive job, a quagmire of sorts. Safe to say, failing earthworks along the crucial stretch of towns and farms between the Portage Diversion and The Forks would not enhance MIT's reputation. Even in a so-called "natural disaster" where responsibility for dikes is distributed among a patchwork of private and public interests, failures resulting from lack of maintenance may well signal to some that MIT engineers are to blame.

On the other hand, the announcement of a precautionary breach of Hoop and Holler epitomizes political engineering at its most sensitive. When engineering becomes most explicitly political, professionals emerge from their usual enclaves and migrate into mass-mediated spotlights. MIT has the power to intentionally endanger the property and livelihoods of those in the breach's inundation zone. (The zone is defined on a probabilistic map that projects Assiniboine responses to the breach. The maps guide preparatory sandbagging efforts.) Sensitivity is intensified by the fact that in 2011, MIT did breach the Assiniboine at Hoop and Holler. This ignites fears, I presume, that not only is the proposed plan a *repetition* of a peculiarly local sacrifice of a rural minority for an urban majority but, too, that the repetition could devolve into a *recursive* pattern, i.e., that MIT might rely on Hoop and Holler every time the Assiniboine floods, just as it relies on the Portage Diversion into Lake Manitoba (a subject of much debate; see chapter 5).

A 2011 logistics analysis identified five possible sites for a controlled breach based on historical flow patterns in the Assiniboine basin. As MIT's (2013b, 79) Technical Review explains, the choice to cut through Hoop and Holler bend, an act that would direct outflow to farms in the *more fertile* south side of the river, involves the following factors:

> The site was chosen as it offered a direct, effective overland flow path to the La Salle River system via Elm Creek. The site also provided for the ability to readily and effectively control flows, easy accessibility via roads, and the greatest degree of protection along the Assiniboine River. The site was located further upstream and so would reduce the flows against a greater length of the Assiniboine River dikes.

Thus, Plan #1, preparing for a controlled breach, helps to avoid negative consequences should the quagmire of Plan #2, walking the dikes to search and plug up vulnerabilities, fail. The law requires that MIT shore up the zone targeted by the breach in the couple of days before the last cut is actually made in the proposed breach. This entails sandbagging and building other earthworks to protect property that cannot be moved out of harm's way. If the final cut winds up having to be made, those harmed will receive financial compensation. How much mitigation is based on the difference between estimates of the actual flood's damage and an estimate of what would have flooded without the proposed breach. Farmers who suffer from the same flood, if not caused artificially, are not legally entitled to financial compensation (see chapters 3 and 4).

As in all instances across the system, definitions of natural and artificial, controlled and uncontrolled, are carefully crafted in law and policy protocols. At the same time, calculations are subsumed by the basic principle that large-scale flooding is a "natural disaster" – a term that the premier articulates at the end of his opening remarks. Technical adherence to that vision requires bracketing off and forgetting about the incalculable effects of urbanization, industrialization, and climate change. In other words, the categorical difference between nature and artifice rolls geoscience-based evidence into a pragmatic solution that is also in part a working delusion (see chapter 5). The maneuver ignores the incalculable nature of the geo-cultural problem and offers officials a rationale protecting them from the full force of the social, economic, and emotional effects of their actions. In practice, no one wants to have to carry out the breach, but in all good conscience, they have to be "at the ready," as MacMahon explains, for the worst, i.e., in technical language "for the upper decile forecast."

III. DISCOURSE ANALYSIS: HOW TO TALK ABOUT TRICKING A TRICKSTER-RIVER

A. Five Semantic Packets

Analyzed as a chronotope, I find five kinds of representations, or semantic packets, of emergency timespace flowing through the communique of officials (see figure 7.1):

1) *past river measurements*: what they knew from measuring how the river acted in 2011

2) *flood forecast*: emerging sense of river action based on continuously updated forecasts

3) *uncertainty*: what they could not anticipate

4) *the known*: what they knew about the river and what they were already doing

5) *action plan*: what they planned to do and why they planned to do it

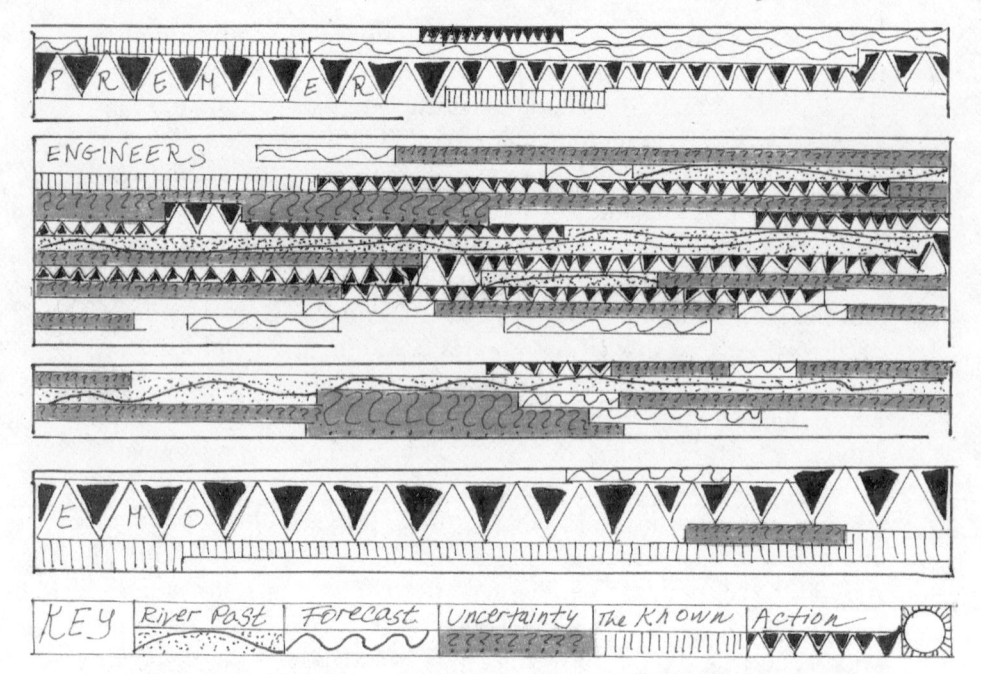

7.1 Chronotope of encounter (emergency press conference). Televised, expert discourse is transcribed (see Appendix III) and coded into visual semantic packets. Action dominates the frame performed by the premier (first speaker) and EMO head (fourth speaker). The shape of uncertainty (shaded areas with ???) emerges from the engineers' mediating discourse (second and third speakers). The engineers hedge questions for which they don't have ready answers between segments that offer substantial geophysical knowledge and that rationalize the proposed breach. This event-specific visual discourse analysis suggests a widely applicable, comparative methodology for studying emergency communication and disaster management. This visual technique works best in colour.

The five narrative segments are not distributed evenly among the officials (see figure 7.1 and the full transcript upon which it is based in Appendix III). The two MIT engineers have responsibility for sketching the geophysical reality insofar as can be known. MacMahon, who speaks the longest and without reading, provides a technical description of the general situation drawing on past river measurements and flood forecasts (semantic packets 1, 2, 4) and layering in caveats about uncertainties (3). He briefly enters the action arena when he explains how they have set a calculated flow "threshold" that would trigger the "last resort" decision to make the final cut in the prepared breach (5).

At the sentence level, MacMahon deftly embeds action plans in the core of his communique. He renders uncertainty manageable by articulating unknowns as precisely as possible and fencing them off with statements about the knowns. He layers doubtful language tactically among solid portions of his communique (historical flood knowledge, last updated forecast, future predicted flow ranges, action plans, rationales). The linguistic effect creates openings for things (geophysical events) to be talked about as maybe happening, without offering any predictions regarding particulars. MacMahon thereby conveys a sense of official knowledgeability, accountability, and transparency despite the uncertainties created by the trickster-river.

In this sense, MacMahon's MIT communique sends uncertainties flowing along defined interfaces between sentient humans and non-sentient river. To convey this aesthetic affect, I composed a textual fragment by identifying and stringing along in original sequence all the words and phrases MacMahon used to connote uncertainty (italics) along with the immediately adjacent certainties that anchor them (bold, with recording minute marker in brackets):

> [I]*ndications that we were going to be looking at flows* **coming into the Portage Reservoir** ... We *anticipate that* **equipment** *is going to be [2:00] arriving* **at the location** *in the very near future.* **And work will commence** *probably tomorrow morning. ... That's not to say we are going to action it. We have to define how the flows come into the site ... We anticipate that* **the Hoop and Holler controlled breach location** *will need to be ready, at the ready* **on Monday or Tuesday** ... **That's our plan** *and we're still on track to achieve that* ... **This is a summer rainfall event** *and the forecasting is a lot more dynamic than* **winter forecast.**

MacMahon articulates two important caveats in this textually uncertain fragment: 1) MIT has to get the province ready for the Assiniboine to do its worst, but if that doesn't happen, then they do not need to and would prefer not to cause a repetition of the hardship and heartbreak of 2011 ("That's not to say we are going to action it."); and 2) summer rainfall events are historically rare, and it is really difficult to predict how they will unfold. In other words, in fact, the river has taken the humans by surprise.

Topping, who follows MacMahon to add forecasting details, adds a third caveat: the *updated and updating forecasts assume ideal weather.*

After the formal presentation, when a questioner asks about the role of rain in the forecast, Topping explains:

> As I said, our forecast is best on ideal weather. We are forecasting due to the high Humidex [heat + humidity index], the possibility of thunderstorms. These thunderstorms tend to be localized, which right now, they're forecasting southeast Manitoba to get thunderstorms, so that would not be a risk. So, it's all dependent on where those storms are centred. If one were to occur over Brandon, it could create problems.

The MIT engineers know a lot of hydrology and hydraulics, but they cannot see into the next few days.[17] Suspense is keyed to the updates emerging from the back offices of forecasters swimming furiously through the digital sensorium. Sudden thunderstorms may skew the system, but where precisely? If they dump to the west over Brandon, trouble; to the east of Winnipeg, crisis is averted. Yet together, the engineers build hedges around uncertainties in the communique. Even if viewers get the gist but miss the details, they come away with some confidence that officials would fairly and pragmatically reinstate environmental security. Or not. Some are angry and frustrated that they sandbagged everything in preparation for a spring flood that didn't come back in 2011, and now this. It is not a happy time for anyone.

EMO Deputy Minister Spencer rounds out the formal performance. His communique is dominated by action. Beginning with a passing reference to the forecast change, he offers the image of soldiers "fanning out with resources to help those identified properties." In his last sentence, however, he includes a fourth and final caveat that encodes a dash of uncertainty: "This will be something done in cooperation with the municipal government and using the military as our backbone, I suppose, to make sure we are able to meet the very tight timelines necessary to protect the property in the matter of three to four days."

Emotionally, the premier, the MIT engineers, and the EMO expert structure anticipation to reduce anxieties and to encourage cooperation. Although fear intensifies for those who may be in the projected inundation zone of the Hoop and Holler breach, I expect that official acknowledgment of harm and existing legal armature for mitigation and compensation helps to quell loss of confidence. The audience understands the basic vulnerabilities. Humans have to wait for a weirdly acting river that may or may not fall into the traps humans have set in place, or set in motion, based on the assumption that when the time comes, upstream skies will clear.

This exchange from the Q & A sums up the tone:

Journalist:[18] This might sound like an odd softball question but this is déjà vu all over again, and I'm just wondering if there's any personal thoughts on how we're going to go through this all over again.
Premier: Well, no one expected to have this kind of an incident three years out since the last one. It's obviously tough on people. We feel for the people who are in the way of the water. We want to make sure they are protected. So, our hearts go out to them. But more importantly, our efforts go out to protect them. So, everyone's gearing up with the same enthusiasm and dedication as they did last time. And I guess we have the advantage of some experience now. We have a clear idea where the water is going to go. The reeves and the volunteers know the task ahead of them. We have better inundation maps now. But clearly, it's a strain on everybody. But clearly, people are in a sense ready.

Others, one woman's voice among them, pose questions about how terms such as cfs might be translated for laypersons, for greater detail about the number of households and farms that will be affected and receive help. After almost twelve minutes of Q & A, the event ends but not before pressing questions about effects on the people along the south end of Lake Manitoba who are again suffering from the use of the Portage Diversion (see chapter 5).

B. Post-Flood Update

Subsequent official reports indicate that things unfold much as MIT forecasters hope. They do not have to breach Hoop and Holler after all. By 10 July, the contingency plan slips from alarming news announced at a press conference to a lesser possibility on the list of important details in the province's Summer Flood Bulletins, of no greater salience than Number-Feet James, which is about to exceed its limit in central Winnipeg.[19] Although provincial cyber-earthworks successfully contain it, the ancient Assiniboine shows itself to be alive and kicking. In July 2014, the river and its tributaries break up a lot of bridges that have been there since the 1920s (among a great deal of other expensive-to-fix infrastructure damage). And although the city of Winnipeg is saved once again when MIT shunted 34,000 cfs (of the total forecast 52,000 cfs) up a refortified Portage Diversion, those waters do not lose their capabilities to retake prairie terrain elsewhere.[20]

And yet, Manitobans are used to this. The Assiniboine flood in summer 2014 faded fairly quickly in the mediascape and looks unlikely to become a significant event organizing popular memory. I suggest, however, that it has left a subtle imprint. The Hoop and Holler families who lived in intense anxiety in the days between the preparation of the breach and the passing of the two river crests surely have not forgotten it. Moreover, it shifts the emotional geography of flooding and flood control toward hazards coming from the west and should continue to raise troubling questions about infrastructural readiness.

Surely flood bowl habitus, that deep layer of dispositions shaping the sociotopographies of security and insecurity, has been disturbed. The disturbance of habitus can only be amplified by knowledge of climate change in Canada and the world. At the very least, most of the experts I talk to believe that climate change is looming over everything, but that doesn't mean that they know what to do. Emergency protocols based on science and engineering tend to rely on recursive patterns to establish norms and rules rather than on interpreting a confusing series of anomalies.[21] In a radically changing climate, the power of expert habitus meets its limit where historical patterns become unreliable. Engineers and architects of built environments will have to come up with more flexible practices that will allow maneuvering according to a wider range of contingencies than those to which they have become accustomed.[22] Climate change may come to bear on geophysical reality faster than forecasters and engineers can change their professional skills and predispositions. Because they are less tied to geophysical contingencies, those with expertise in arts and culture may find it within their ambit to inspire a shift in geological imagination strong enough to remold habitus into a new blend and set the stage for a turn in engineering practice. Who knows?

In any case, after analyzing the emergency press conference of 2014, I find myself impressed with its balanced, informative tone. Quite civilized, actually. If I were a Winnipeg inhabitant afraid of an approaching flood, I might want a team like that looking out for my neighbourhood. MIT engineers MacMahon and Topping prepare the populace for the possibility of a controlled breach. They couch their traumatic message with forecasts that help to interpret the need for it, while presenting Assiniboine's uncertainties with humility. Framing the engineers' communique, the premier and head of Emergency Measures convey a sense of cautious mastery governed by respect. The argument that it might have been better for the majority to sacrifice a small infrastructural outsider-minority in a controlled breach at Hoop and Holler rather than risk an uncontrolled one appears to have been well-supported. In that way, control is expertly performed in the media and in the river.

IV. STRUCTURES OF FEELING: FLOODS IN LITERATURE AND THEATRE

Putting Bakhtin into practice in a mass-mediated governmental context, I analyze the press conference as a "chronotope of encounter" (1981, 243) that transfigures the Assiniboine and assimilates it into a tamed, though still uncertain, televised timespace.[23] Experts sit on a televised stage, facing off-camera journalists, while socially diverse inhabitants come to their screens – together, they encounter a threatening-but-well-studied Assiniboine-in-flood. Even in the focused TV timespace of encounter, the trickster-river may cause anything to happen. Fear of a river let loose haunts the event and lends officialdom intensity; the fear crosses generations through memories passed down through time in photographs, stories, and system focal points like Number-Feet James. Yet fear is also contained, geophysically by infrastructures or distance and culturally by the official performance of hydrogeological data, i.e., by enactment of engineering expertise, geoscience knowledge, and political authority. Tension between fear and confidence characterizes the "structure of feeling" that emerges when water threatens to fully enact its elemental powers in a city well prepared for disaster.[24]

To explore the feeling-tone of fear that haunts confidence in provincial control and contributes to structures of feeling that recursively emerge with extreme floods, I turn to literature and theatre (Bakhtin's familiar terrain). Moving out of flood engineering into flood fiction, the chronotope of encounter combines with what Bakhtin (1981, 243) calls the chronotope of threshold. In these fictional threshold moments, flood timescapes relevant to populations merge into the background as plots condense around thresholds experienced social-interactionally and personally. Crisis manifests emotionally as the narrative dramatizes a complete break with normalcy and with biography.[25] These art forms capture the feeling-tone of experiencing the river's agency at the point of being overpowered – when the human sense of duration stops, a moment of utter mystery.

A. *William Faulkner's* The Old Man

The fragment of high art I import first takes place during a flood in the Mississippi River, another tentacle created by the Outburst Quartet. The selection comes from William Faulkner's (1948, 48–9) novel *The Old Man: Violence and Terror in a Mississippi Flood*. The old man *is* the Mississippi River. It is 1927; a convict released from prison to fight the flood finds himself in a small boat with a very pregnant woman he has rescued from a tree. This is a literary master's recreation of what an uncontrolled breach feels like:

> So he was paddling on, strongly and steadily, not alarmed and not concerned but just exasperated because he had not yet begun to see any reflection on the clouds which would indicate the city or cities which he believed he was approaching but which were actually now miles behind him, when he heard a sound. He did not know what it was because he had never heard it before and he would never be expected to hear such again since it is not given to every man to hear such at all and to none to hear it more than once in his life. And he was not alarmed now either because there was not time, for although the visibility ahead, for all its clarity, did not extend very far, yet in the next instant to the hearing he was also seeing something such as he had never seen before. This was that the sharp line where the phosphorescent water met the darkness was now about ten feet higher than it had been an instant before and that it was curled forward upon itself like a sheet of dough being rolled out for a pudding. It reared, stooping: the crest of it swirled like the mane of a galloping horse and, phosphorescent too, fretted and flickered like fire. And while the woman huddled in the bows, aware or not aware the convict did not know which, he (the convict), his swollen and blood-streaked

face gaped in an expression of aghast and incredulous amazement, continued to paddle directly into it. Again he simply had not had time to order his rhythm-hypnotized muscles to cease. He continued to paddle though the skiff had ceased to move forward at all but seemed to be hanging in space while the paddle still reached thrust recovered and reached again; now instead of space the skiff became abruptly surrounded by a welter of fleeing debris – planks, small buildings, the bodies of drowned yet antic animals, entire trees leaping and diving like porpoises above which the skiff seemed to hover in weightless and airy indecision like a bird above a fleeing countryside, undecided where to light or whether to light at all, while the convict squatted in it still going through the motions of paddling, waiting for an opportunity to scream. He never found it.

Most of Faulkner's novel takes place in the aftermath of this disorienting encounter. During a temporary respite, a warning that a hole will be breached in the levee passes among his Cajun companions. But he doesn't speak French, so he does not understand. His companions help to save his life by forcing him to get on their boat. He persuades them to bring along his skiff as they escape. The struggle goes on and on, but in the end, the convict manages to bring the woman and newborn back to the vicinity in which he first found her. After saving them, he is relieved to make it back to prison with another ten years on his sentence for purportedly trying to escape. He tells his rapt fellows the tale of his marvelous encounter with the Mississippi, and there the novel ends.[26]

Next, I turn to theatre, another literary genre that mobilizes the aesthetic power of flood trauma, where the Red River, that south-to-north running tentacular remnant of the Outburst Quartet and Winnipeg's traditionally greatest threat, brings the stage to life.

B. Ian Ross's The Gap *and Tennessee Williams's* Kingdom of Earth

Theatre provides a live stage for performing geo-culture in and about Winnipeg. In *The Gap*, Métis/Ojibway playwright Ian Ross (2001, 11, 109) crafts dramatic action in the timespace of an unfolding flood. Drawing on water's destabilizing potential, Ross drives suspense into a plot of cross-cultural human encounter that is at once universal and unique to Manitoba. Unlike the informative yet reassuring tones of the emergency press conference, *The Gap* brings the audience into the anxious timespace of oncoming flood retrospectively, at some distance from actual geophysical danger. This gives the playwright freedom to charge characters' encounters with the compulsive impulses of rivers in the city.

In so doing, I suggest, theatre arts enter into a sphere of unintended agencies, indirectly crossing paths with off-stage engineers and officials who, although they have no stage presence, tend to the system that offers the characters their relative sense of security. In his play, Ross brings the theatre into the "semiotic commons" of flood control, contributing to "a social infrastructure of communicative channels" (cf. Elyachar 2010, 452).[27] Of course, theatre makes meaning quite differently from shared reference points like Number-Feet James, a communication tactic and civic infrastructure that actually organizes survival in flood emergencies (see chapter 3). But *The Gap* and Number-Feet James both integrate shared reference points; they both occupy and give meaning to the chaotic spaces of Winnipeg's flood history.[28] The semiotic commons crosses the spectrum of experience and reflection, integrating joyful as well as traumatic relationships with the rivers.

Unlike the fragment of Faulkner's novel, which is primarily a chronotope of threshold (the characters are inside the river's crest), Ian Ross imports a timescape similar to the press conference. In his chronotope of encounter, the characters are watching and waiting for an oncoming river crest that may or may not arrive. Ross's play is actually quite an apt example of encounter in Bakhtin's (1981, 243) original conception, more so than the press conference, because the play explicitly makes "paths of the most varied people – representatives of all social classes, estates, religions, nationalities, ages – intersect at one spatial and temporal point. People who are normally kept separate by social and spatial distance can accidentally meet; any contrast may crop up, the most various fates may collide and interweave with one another." The scale of social interaction in *The Gap* is intimate. While Ross does not reach for the carnivalesque possibilities of Bakhtinian encounter, his play does capture the mystery of people who find meaningful relationship across entrenched settler-Métis boundaries.

When performed in Winnipeg's Prairie Theatre Exchange in 2001, Ross must have brought many in his audience back into their experience of the most recent major flood, the flood of 1997 that put the modern control system to the test (see chapters 3 and 4).[29] The control system succeeds in protecting the city's inhabitants but not without suspense. The Floodway was just one rain away from disaster. His characters enact, and are accidently empowered by, the ambivalent yet hopeful mood of appreciation in this moment of the city's flood history. Sandbagged landscapes set the stage for the approaching deluge on the Red River. Everyone hopes that the Floodway and supporting earthworks will hold.

Act One opens on the riverbank with this stage direction:

Lights up on an odd landscape. It very much resembles a dry riverbed, only magnified. Cracks and gaps create a disjointed mosaic of earth. Several sandbags are present; they form a small "wall" that cuts across the middle of the stage. A well-dressed, middle-aged woman walks on, VI CHAMBERLAIN. *In her hand is a leash, attached to a small dog,* MR. CONNELLY...

The "cracks and gaps" of the riverbed represent the cracks and gaps among the Winnipeggers who share a city but not everyday lives. The earth and the people await the deluge. The sandbags establish a literal wall that divides and connects the movements of characters within and between the interiors of two family homes. As metaphor, the approaching flood serves multiple functions. It sets the time span in which the river crest approaches, framing dramatic action. Like the character of the dog, the approaching flood elicits different levels of concern and behaviour. The differences help to flesh out the personalities and cultural identities of the characters.

Vi's dog is anxious, and she, distracted and chewing on her nails, tells the animal: "Don't be frightened. The flood won't touch us. It won't." While she refuses to engage in a practical way with the threat, volunteers arrive to help sandbag her home. As she exits the stage, family members and volunteers come on. Upon entering, the elder character, Saul, watches the river and recalls the last great flood. In this, the playwright explicitly imports one flood from the habitus of memory. But in his next stage direction, he also sets up the encounter chronotope, that is, the racial and class identities that structure tensions across city space in times of crises and normalcy:

A middle-aged Métis man, SAUL MACKAY, *and his youngest son,* CHESTER MACKAY, *enter.* SAUL *shakes his head and clucks his teeth.*

The men are there to help with sandbagging the house. Their presence sets up the trope of community common to almost all historical flood narratives: in disaster, everyone sandbags together, overcoming any and all barriers of class and race. But Ross writes the dialogue between father and son so as to capture the resentment that may arise through the cracks and gaps of volunteer sandbagging. Though spoken of in conversation (including in my interviews with non-Indigenous folk), such resentments are rarely put forward in public or in published accounts. The play also narrates against the stereotype that Indigenous people are not also city people[30] and, too, that the divide between infrastructural outsiders harmed by flood control and the protected majorities do not map neatly onto social categories such as Indigenous–non-Indigenous. The play's fictional register can thus more accurately portray reality than much canonical history and folklore.

> SAUL: Look it all that water. I ain't seen that much since 1950.
> CHESTER: Let's go, Dad.
> SAUL: What the hell for?
> CHESTER: I feel weird.
> SAUL: What do you mean, you feel weird?
> CHESTER: I feel weird walkin' around white people's houses and yards. What if they think I'm trying to break in?
> SAUL: Chester.
> CHESTER: What?
> SAUL: There's some of their blood in you, you know.
> CHESTER: I know, but I don't care about that, I know if I saw some white guys wandering around our house I'd think they were tryin' to break in.
> SAUL: What are you getting' all paranoid about? If white guys were tryin' to break into our house I'd call the cops.
> CHESTER: If white guys were tryin' to break into our house they'd be the cops. Coming to arrest us.
> SAUL: Quit talkin' stupid. Make yourself useful.
> CHESTER: Where's Evan?
> SAUL: Never mind Evan. Just start sandbaggin'. Would you look at all that water? Not since 1950. But you know what? Chester. You know what?
> CHESTER: What?
> SAUL: Some good's gonna come out of this.

Saul's remark, "Some good's gonna come out of this," resonates with an old geo-cultural theme central to Dutch (now globalized) national identity – floods are an equalizing force because they compel cooperative action across social boundaries. In other words, "some good's" come from centuries of surviving and thriving dangerously close to water. In the Netherlands, long before government, a fundamental good arises from water management – the very idea of democracy.[31] And concomitantly, out of Dutch geo-culture arises a technological prowess with global reach that extends into Manitoba. The Floodway gates, in fact, are built with the help of a Dutch technical expert.[32]

Within this far-flung techno-cultural web, the play unfolding in the Floodway's protected zone proceeds:

Vi's daughter enters. Her actions will initiate the crossing of class and race divide through the medium of a romantic relationship with Saul's other son Evan. The plot revolves around how the lovers, and the two families, recognize and negotiate cultural differences. The good that Saul predicts in Act One does come to pass in two ways. The rapprochement between these two families across the cultural divide is a good thing. But more deeply, the good also lies in the way that the new relationships help each character to realize the mixtures of racial identities and suppressed stereotypes that compose their sense of themselves and others. In this, the plot speaks to a founding and persistent problem in Canadian cultural politics, one that inhabitants may have been hesitant to speak about back in 2001 when the play was first performed.

At script's end, Ross explicitly merges the fear of floodwater with the fear of flooding emotion. The dialogue closes on the two lovers. Vi's daughter says she can't say how she feels about Evan:

DAWN: Because ... if I let you in ... into all of me ... I'll be flooded.
EVAN: ... Some floods can be good.
Lights out.
The End.

Loss of terrain, of coherent boundaries that define one's sense of self is here projected onto the loss of terrain to the river. As an ethnographer gleaning flood content from Ross's script, I am disappointed but not surprised that the action and dialogue is not particularly focused on flooding or flood control. I compare it with a performance of *Kingdom of Earth* by Tennessee Williams that I see in St Luke's Theater of New York City in the summer of 2016. This classic American play also takes place in the Mississippi Delta, this time in the infamous flood of 1950, the disaster that followed the flood of 1927. Three desperate characters wait to either die or climb up on the roof with the chickens. They listen to the river's roar, the sound of God, or, as the character corrects himself as he peers out the door of the house to say, the sound of the neighbour who is about to blow up the levee, an act that would protect the neighbour's place while destroying those around him. The play ends at the threshold, the murderous sound of the explosion. We are left not knowing the outcome of this breach made without the kind of legal protections that would probably stop a person from carrying out such a dastardly deed in contemporary Manitoba.

Like Ian Ross's *The Gap*, Tennessee Williams's play is a chronotope of encounter. The flood metaphor organizes the temporal frame of action. The plot itself revolves around an intensely emotional life and death negotiation of sexual desire and power among two brothers and one woman, a stranger to the brother living in the house. The similarities in the way both Ian Ross and Tennessee Williams use the flood metaphor without making it central to the plot leads me to push the analysis further. I realize that in addition to motivating plot and character, the flood metaphor projects an intensified affect, an aura of dreadful anticipation, to everything that transpires. The suspense of an approaching flood is a state-of-being that playwrights borrow from the well of audience memory to propel emotion into a written script and on to the performance stage. On-stage suspense mimics history so well that audiences can ride upon their memories (of experienced events, films, newsreels, Faulkner's novel, etc.) to enter into the characters and action. Suspense originates in the real and is transportable. Feeling-tones enrich blended habitus. Through aesthetic power, the arts indirectly support and enact flood control.[33]

V. THE SHAPE OF UNCERTAINTY

The uncertainty of riverine agency drives the suspense of flood encounters. Across contexts and genres, whether official action in the midst of emergency (as in the press conference) or fictional personal experience in past emergencies (as in the novel and plays), uncertainty is the unifying motif. The selections here can be arranged on a scale of greatest to least narrative uncertainty, with Faulkner's heroic convict story representing the greatest. Recall his situation in the threshold: "the skiff seemed to hover in weightless and airy indecision like a bird above a fleeing countryside, undecided where to light or whether to light at all." In contrast, provincial press conference officials expertly resist as they articulate underlying uncertainty, thereby producing a TV narrative that emanates a sense of being protected from the trickster-river. Recall Chief Engineer MacMahon, who I find conveys a sense of official knowledgeability, accountability, and transparency despite uncertainty.

Now I want to turn this around. What if political engineering in general, and the press conference in particular, are not about communicating a plan of action *despite* uncertainty? What if it is, instead, about communicating the *shape of* uncertainty? Looking again at the textual fragment (the indented paragraph on p. 167), consider how the shape of uncertainty emerges from that which is certain, literally. Visual insight emerges from a reverse analytic here. Rather than highlighting the sureties, a perceptual twist allows the italicized phrases that index uncertainty to emerge as a shape, a semantic field, an active conceptual entity in and of itself. This reverse visual analytic offers a path of knowledge into the sphere of unintended agencies. I suggest that the reverse analytic captures what people are unconsciously sensing; it visualizes an unarticulated aspect of discourse that can produce a calming affective response among the majority population. By performing politically, engineers show that what they know about the river city goes beyond the assurance of knowable facts. Unlike Faulkner's bewildered but doggedly rowing hero-convict, the televised political engineering cadre assume a collective stance in which riverine uncertainties are qualified and doubts informed.

The city is not and will not be overwhelmed at the threshold of the Assiniboine crest in 2014. No. Provincial experts are prepared. They are calculating flows coming through the Portage Reservoir; they have equipment and soldiers, and they know where they will send them; they know the hydrological threshold that, if it must, will trigger the last cut of the breach; they know exactly where this will be and have specified a two-day window; and they recognize the complications of a summer rainfall emergency. This is evidence-based action at its best – not without uncertainties but, rather, with defined uncertainties. Each – the known and the unknown – depends on the other to demonstrate its effect.[34]

And so Manitoba's most technically sophisticated flood control engineering practices send feelers into the sphere of unintended agencies where the Assiniboine rises, pulsing through the delta to tangle with the city set on the home base of Lake Agassiz's ghost. Ultimately, it is the shape of uncertainty, I argue, that distinguishes the discursive capabilities of political engineering.

8

Avulsion!

I. THE ADVENT OF PLURILINEAR POSSIBILITIES

Manitobans live in tension with water's impulsive, implacable force. Cities enact human collectivity by reshaping themselves around their flood-prone rivers; rivers enact riverhood by occasionally rising up out of familiar channels to submerge the city. This is the framework of the book, and I have confidence in it. But the time has come to disrupt these imaginative underpinnings. In this last take on the geo-cultural identity of river-city beings-in-tension, I enter into a more radical source of geomorphological uncertainty. Avulsion!

What if flood-prone rivers are haunted by deep time ghosts with the potential to seduce them away from historically familiar channels, paleochannels, that repossess their waters if you will or that welcome them home to reinvent themselves as live actors in a most improbable geological future? Impulses of rivers and cities combine as they carry on being and becoming, persisting in distinctive, yet entwined, embodied form as is my premise; but what happens, I ask now, if geophysics makes persistent distinctions impossible?[1] What happens when expert-inhabitant coalitions come to rely on a mixed version of natural order in the world region they know best and it gets suddenly sideswiped by an alternative version of things? This is the crux of the climate crisis, is it not?

The queer behaviour of the Assiniboine River in the floods of 2011 and 2014, Winnipeg's last two major floods, revealed its talent for issuing climate change communiques.[2] So it should be no surprise that the Assiniboine has another trick up its basin. Its longevity and long, meandering course are qualities that lend the Assiniboine the power to smuggle a family of alluvial ghosts into the techno-savvy house of engineering. Perched above the prairies, the lower reaches of the Assiniboine hold the power of avulsion. An extreme act of riverine unpredictability, an avulsing Assiniboine could confound all norms, expectations, and built environments developed since the city of Winnipeg's beginning.

The meaning of avulsion travels in law and geopolitics,[3] but geomorphologically, avulsion refers to channel migration of a particular kind (Bierman and Montgomery 2014, 193):

> In addition to gradual channel migration by cutbank erosion and point-bar deposition,[4] *channels may **suddenly move to a new position**, a process called avulsion* ... In addition to meander cutoff events, channels sometimes avulse when a flow obstruction like a logjam blocks the channel and causes flow to spill over the streambank and carve a new channel across the floodplain *or shift flow into a secondary side channel or to an inactive abandoned channel.* [my emphasis]

Whenever there's a major flood on the Assiniboine, there's a danger that the river might burst out of its channel in an uncontrolled way. To avoid this, MIT carries out a controlled breach in 2011 and prepares to do another should it become necessary in 2014. A river leaping out of its channel and flooding out neighbourhoods and farms willy-nilly is nothing to chuckle about. Avulsion, however, ups the ante. To be clear, the last act of Assiniboine avulsion puts the river where it is now, where it has been long before settlers arrive.

These days, city boosters perform some sleight-of-hand when it comes to representing the geo-cultural significance of The Forks. To tell a good story about why it is so important, they represent it as the "ancient" confluence where Indigenous people gathered about 6,000 years ago. However, "There is yet another story to be told," Bill Rannie tells me as we look out at the alluvial fan west of the city.[5] Archeological excavations find evidence of habitation going back only 3,000 years or less, and even then, Indigenous people probably met at The Forks periodically. They don't establish sedentary camps there. About 3,000 years ago, the place inhabitants call The Forks is located not in central but in south Winnipeg, closer to St Norbert, at the present-day confluence of the La Salle River, a living paleochannel, and the Red. The narrative sleight-of-hand seems to be a relatively benign example of what Jeremy Schmidt (2020) calls "settler geology," wherein settlers mine geology for the explanations and timescales that suit their purposes, which often distort Indigenous peoples' territories and histories.[6]

In any case, this is the main point: if the Assiniboine avulses again in the near future, it could endanger a city with a population of about 700,000. It could avulse into the La Salle in an uncontrolled way, sweeping into Winnipeg north of the Floodway. Actually, the Assiniboine could seek out and flow into any one of a number of the other waiting ghost channels. As a result of this improbable occurrence, Winnipeg may no longer look anything like today's Winnipeg. Even if there might be flood control engineers capable of preparing for a subset of contingencies, Assiniboine avulsion could potentially disrupt the city's basic spatial organization.

If an act of avulsion in geological time intersects with historical time, it would undermine the empirical bases for the built environment and the carefully crafted alliances of law and engineering that sustain it. So, although the Red is way bigger than the Assiniboine and it's the Red that repeatedly morphs from river-size to metaphorical sea-size, the Red, at least on the Canadian side of the border, has been largely tamed by an adaptable modern flood control system. It seems to me that it is trickster Assiniboine that is more likely to evade every conventional practice of flood control security.

Knowing about something so potentially dangerous should lead experts and inhabitants to contemplate possibilities deep in the sphere of unintended agencies. For example, depending on which direction the Assiniboine goes in, seeking out a new paleochannel to flow into could make moot the hierarchy of operational rules that achieve system balance through river diversion and lake level manipulation (see chapters 3 and 5). For people along the internal frontier in Lake St Martin, the upheaval could ultimately be a relief rather than a problem because it would open up the possibility of a more just arrangement. And so, like the pelicans enjoying the great fishing spot on the side of the Portage Spillway, there's always a chance that avulsion might make everything topsy-turvy in a good way (see chapter 1).

A. Bathtub Experimentation and LiDAR Visualization

How might geoscientists and engineers predict if and when a river in a hydrological system might avulse? Step momentarily into the laboratory of Doug Jerolmack (2014) and group, experimental geophysicists interested in the Atchafalaya-Mississippi flood of 2011, a record-breaking event in the remnant north-to-south tentacle of the Outburst Quartet. To get a better understanding of deltaic processes, they set up a simple system to simulate deep time. Adding a steady flow of water and sediment with a narrow range of particle sizes, they are able to observe "incredible spontaneous complexity" in a bathtub. Watching and measuring their model river's tendencies to move toward places it hadn't gone for a while, drawn by steeper slopes, the group can predict the average rate of channel jumping.

Writing up the bathtub experiment for an interdisciplinary audience, Jerolmack (2014, 81) ends his essay by firmly rejecting the fantasy that experimental knowledge can ever give engineers the power of river control, but he does offer probability. And thus, by producing experimental calculations with universal implications, geoscience gives shape to uncertainty:

> There is a mass balance for the system, overall, which we can measure and understand. Yet, in natural systems, there are things that are predictable and things that simply aren't. We can never predict with certainty, despite our absolute best efforts at control, where the river is going to go, and when it's going to go there. All we have is probability.

Data collection expands out of the laboratory and into the atmosphere above the floodplains: Like physical experiments, visual technologies produce knowledge that gives shape to uncertainties. Each new geo-visual technology has the potential to push geo-culture in a new direction. For example, in MIT's (2013b, 13–21, 91–2, 113–14) review of the 2011 flood, LiDAR, a new remote sensing technology, adds its computational laser-pulsing precision to the earlier knowledge produced using traditional aerial photography, physical experiment, and hand-drawn maps.[7] The intergenerational data compilation shows the Assiniboine finding the apex of the softer alluvial fan and bursting free from the confinement of glaciofluvial drift to become a pluri-linear body (see figure 8.1).

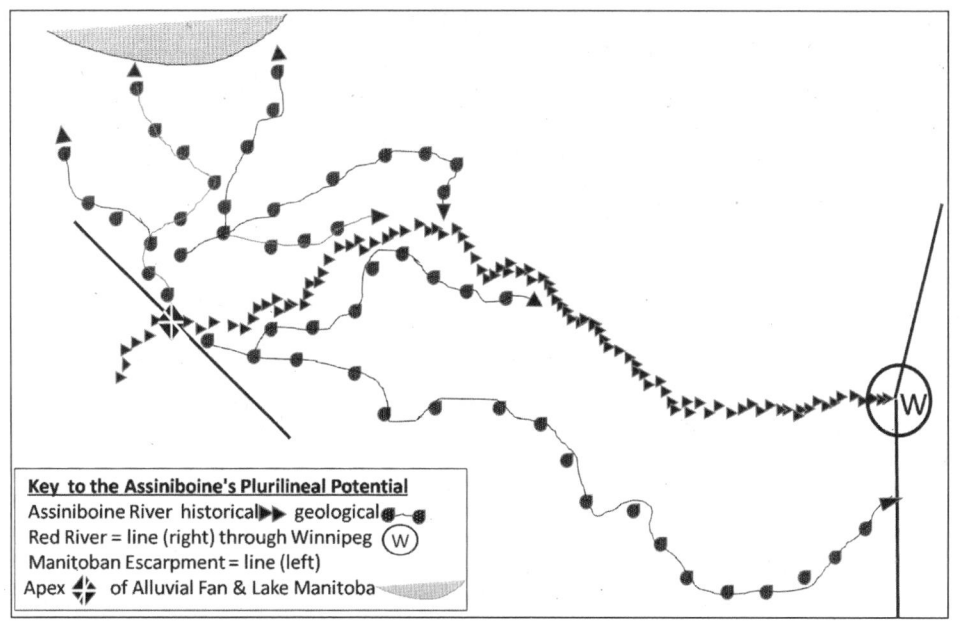

Key to the Assiniboine's Plurilineal Potential
Assiniboine River historical ▶▶ geological
Red River = line (right) through Winnipeg (W)
Manitoban Escarpment = line (left)
Apex of Alluvial Fan & Lake Manitoba

8.1 Assiniboine avulsions. From the apex, paleochannels fan out in the delta space between Lake Manitoba and the Red River (arrows = flow direction). Heading directly into downtown Winnipeg, the Assiniboine's historical channel is haunted by this deep time meshwork revealed by LiDAR technology. Climate change adds significance to the river's plurilinear potential.

Figure 8.1 is an instantiation of an empirical data set that builds theoretical understanding across generations. But it is best understood as a mobile representational element in a set of engineering, policy, and educational practices that constitutes the known river. In this sense, the drawing works in tandem with other modes of place-based communication to perform geo-culture. For example, as he has many times in his long career, Rannie shares his intimate knowledge of the Assiniboine's plurilinear being with his geography students (and the ethnographer) on a sunny 21 September 2014:

> As we set out on the Geomorphology Field Trip, Bill Rannie hands each of us a twenty-two-page handout with aerial and ground-level photographs, maps, graphs, charts, diagrams, data on historical floods, infrastructures, soils, and more. The delta created out of the encounter of the Assiniboine River and Lake Agassiz is the first of many sites we visit as we drive west and south, stop-

ping at river's edge for lunch and at Tim Hortons, of course, for coffee before looping back to Winnipeg. As we climb out of the vehicle at each site, we follow Rannie to a vantage point where he talks and points. Linking handout page to landscape feature, from the vast Pembina valley to the tiny swallows' nests built into an aeolian (wind-borne) sandbank, Rannie decodes the prairie. His revelatory performances transform the mundane into science-magic, the arcane into fun. True, with the benefit of time and technologies, understanding of ice age end has come a long way since Upham's nineteenth-century travels by horse and wagon (see chapter 2). But the sense of discovering the deep time powers and patterns of rivers persists.

Da Cunha (2019, 251) captures the geo-cultural implications of such educational scenes: "With a river system and city in place, facts are not just read differently; they are constituted differently." And we can extend Annemarie Mol's (2002) ethnographic theory of a Dutch hospital, where she discovered that each technology-based set of practices creates a different sense or version of the body – in this case, of the Assiniboine River's body. Once the interpretive frame expands to capture the intersection of historical and geological time, all versions of the Assiniboine are part of empirical reality. And thus, the bathtub experiment, the LiDAR image, and Rannie's geomorphology field class together reveal the empirical river to be in fact, a "body multiple."[8]

What visual trick communicates the intersection of historical and geological time to (social) scientists, their students, and other inhabitants? LiDAR images reveal that in the past 7,000 years, what we know of as today's river has flowed through eight different channels. Yet the image itself shows Assiniboine's multiple bodies to be co-occurring. Like Teller, Leverington, and Mann's (2002) geovisual model of the Outburst Quartet, the timescapes of the body multiple collapse geological phases into a singular landscape array. The array represents remnants of past events and, like Jerolmack's experiment, probabilities. Only one of these lines, now perched well above the prairie, conforms to the Assiniboine familiar to inhabitants.

For those paying attention, like Manitoban hydrologists and MIT engineers and policy experts, the Lidaresque probabilities of Assiniboine's body multiple must, I assume, destabilize the conventional sense of how geophysical reality determines everyday life. I imagine it's as if someone who lives nowhere near California suddenly discovers that the story about their city lying on a San Andreas–type seismic fault is actually too true. And while those who live and farm in today's fertile alluvial fan might notice variations in soil type, given that over the past 7,000 years or so all but the LaSalle channel have been filled in with sediment, they might easily not be aware of avulsion as a phenomenon relevant to their lives.

II. RIVER DIVINATION

This little theory is tentative and could be abandoned at any time.
Robert Smithson ([1968]1996)
in "A Provisional Theory of Non-Sites"[9]

And so, Manitoba's flood control system is built on a unilinear sense of river-city settlement history, past-to-present-to-future. But while the Assiniboine flows toward the Red and Winnipeg in a singular channel, in the context of geological events its unilinear character reveals itself to be an artifact of humanity's comparatively short existence. In contrast to the more powerful and predictable postglacial Red River, one could say that the preglacial Assiniboine is a "little theory that could be abandoned at any time."[10] No matter how successful MIT's flood control system, we know its provisional nature in more detail than ever before.

A question: When Manitoban engineers redesign systems and operations for climate crises, should they abandon modern theory of flood control for another, or others, that better theorize the multiplicity of geological river-being? Does it make sense and is it feasible to prepare for the plurilinear possibilities of geological time? Engineers might think of LiDAR's evidence as an artifact that presents a tentative theory and go on with things as they are. Or they might marshal forces to prepare for the possibility that presents the clearest danger by, for example, mobilizing supplementary resources, both existing and invented to suit the circumstances.[11]

Geovisual knowledge gives shape to uncertainties, yes. Sometimes, though, it can significantly *add* to the uncertainties to which they give shape. Engineers at MIT (2013b, 20–1) have certainly given thought to what could happen in light of LiDAR's revelations. The possibility of avulsion has direct implications for controlling overbank flows during large floods. If the Assiniboine breaches its banks, overflows could:

- flow away from its current channel, never to return (under unregulated conditions) and
- follow paleochannels that ...
    - flow toward Lake Manitoba, increasing flood risk there
    - arc north and then back east into the Assiniboine, threatening Winnipeg
    - arc south into the La Salle and then into the Red inside Winnipeg, inside the Floodway
    - have unknown or less obviously dire effects[12]
- or create a new avulsion to a lower but unpredictable location

Implicit in the MIT analysis is a three-phase sequence: 1) floodwaters tearing away from channel; 2) overland flows scouting for alternative channel opportunities; and 3) flowing into one or more viable channel alternatives. The focus of current engineering action is the prevention of Phase 1, tearing away. The conclusion of MIT's (2013b, 112–13) *Technical Review* finds that with regard to avulsion:

> The flood of 2011 was destructive, causing significant damage and hardship. However, if the Assiniboine River had abandoned its present channel and followed a new course, the implications would have been far more significant and longer lasting. Thus, the water control works not only provide flood protection, they also provide *a measure of stability*, which helps to maintain the Assiniboine River's course *in a naturally unstable and dynamic reach* of the river. [my italics][13]

Flooding is usually talked about as a problem of excess water, as in the above paragraph. But avulsion also risks loss, a lack of water not driven by drought. What if, as a result of an extreme flood, the Assiniboine goes elsewhere? To borrow from Philippopoulos-Mihalopoulos (2015, 8), what if it withdraws and in so doing transforms the prairie wetland assemblage it has sustained together with other bodies throughout human history, oral and written?

How would city life change without the Assiniboine? The Forks would not be a confluence; it would just be a spot along the Red River. The focal point at Number-Feet James that coordinates expert and inhabitant communication in flood extremes would become irrelevant. The old Assiniboine channel could be left to restore itself as prairie or forest, maintained as a park that doubles as a potential water storage site, or filled and paved over for roads and real estate development. Assiniboine avulsion would have to compel another historic shift in blended habitus, a shift as significant as that initiated by nineteenth-century paper survey maps of the land around The Forks that first impose the Dominion's sedentary settlement pattern, dispossessing First Nations, and also a shift as potent as the twentieth-century infrastructural system, which provides safety for the densely populated city folk while harming those who are transformed by spatial location into infrastructural outsiders. In the twenty-first century, the possibility of an Assiniboine act of withdrawal-by-avulsion could open up a chance for humans to remix blended habitus in a way that fully centres social and environmental justice in the alliance of engineering and law.

When it comes to public discussion of flood control operations, Winnipeg inhabitants funnel a great deal of attention to the fraught politics related to Portage Diversion operations. However, to my knowledge, thinking about how avulsion might impact central Winnipeg remains, except among some experts, in the realm of what Bourdieu (1985, 168) would call the universe of the "undiscussed (undisputed) within a larger universe of discourse (or argument)." But the phenomenon need not remain obscure. Combined with related openings in public discourse afforded by the climate crisis, the idea of Assiniboine avulsion has the potential, I believe, to open up a panorama of productive conversations and realignments among expert-inhabitants and river infrastructure.

Geovisualization of Assiniboine avulsion illuminates the phenomenally complex nature of river-city entanglements and the curious instability in its state-of-being. Assiniboine's avulsions suggest another way to conceptualize how its river runs along earth's surface. Rather than an engineered mono-linear channel with land on either side, the river is a plurilinear meshwork with dry land and wetland spaces between channels.[14] At present, only one of Assiniboine's plurilinear possibilities is powerfully integrated into the flood control technozone. The central pragmatic question is still, how are engineers to divine the infrastructural future now that LiDAR details such a different hydrogeological ordering principle?[15] Assiniboine's abandoned channels register possible futures as well as pasts. The choice is not between a singular path and everything-is-everything relatedness. That would stymie thinking and action. No, a distinct shape emerges – the shape of uncertainty defined by a new visual technology, a shape that calls for a new engineering vision.

A. Sustainability of Which?

There is an emerging global consensus among urban planning experts that rivers should be given more room to filter into and meander through their floodplains, both inside and outside cities.[16] I agree, within spacetimes of human histories, it would surely be more sustainable to move buildings away from banks, let a river swell, shrink, and meander across a larger swath of floodplain.[17] But avulsion, it seems to me, presents a dilemma of a different kind and scale. Thinking in terms of avulsion is not about giving a river room to meander or to braid around a wider but still-unilinear zone around a channel. Rather, to account for avulsion's possibilities, expert-inhabitants might, perhaps, begin to think about how to materialize infrastructures that ramify into a meshwork of different floodplains.

Might engineers – who come from different backgrounds and may include Indigenous people – be able to adapt their considerable technological powers of observation, analysis, geophysical interventions, and hierarchical decision-making procedures to figure out how to accommodate more flexible, mobile, precautionary practices? Might they integrate some practices consistent with Indigenous forms of blended habitus – which do not tend to assume historical linearity – and thereby begin to think in ways that give more room and agency to rivers?[18] The new version of the technozoned river city could be part of a meshwork, out of which a new blend of habitus might emerge, that 1) acknowledges both geophysical and technological limits and uncertainties; and 2) better distributes the spatial injustices of modern flood control infrastructure.[19]

In other words, the idea of giving rivers room to flow so as to sustain the configuration of the city makes sense, but it does not begin to grapple with an improbable but highly consequential avulsion event. As I define it, flood control is an intentional collective act to keep the city dry, an effort to invent and to sustain an altered state-of-being shared with rivers. The recently acquired knowledge of Assiniboine's avulsion powers challenges the very idea of sustainability – which sustainability?[20] The emphasis of flood control has to shift to a conceptual space that aspires to sustainability, invention, and legal pluralism in geological as well as historical timespace.

And too, while avulsion may be improbable – timed by geological, not historical, rhythms – the danger of the Anthropocene points directly to increasing occurrence of exactly such hazardous intersectional timescapes. Geoscience predicts that record-breaking flood events will probably increase in intensity and frequency in the climate crisis (Caretta et al. 2022). This will increase the probability of avulsion. In other words, avulsion throws wild speculation into the geo-cultural mix, but it is not science *fiction*.

But knowledge coded in probabilities is a funny thing. Do the LiDAR images provide too many probabilities or too few? In theory, forecasters might be able to inform engineers if and when the Assiniboine might avulse by observation and measurement of an intense overland flow edge moving too close to a paleochannel. In practice, overland flows are hard to measure and predict, complicated by roads, railway embankments, and culverts. And by the point in the forecasting process when the river crests, bursts out, and when flow edges appear to be approaching a paleochannel, there's always too much data streaming in and rarely enough time to interpret and act on it (see chapter 3).

LiDAR reveals Assiniboine avulsion to Manitoban engineers in a historical moment that may compel speculation. The northern reaches of the continent wobble between flow patterns created in the tentacular outbursts of the last climate change, when glaciers melt and retreat to the pole, and the present climate change, when the polar ice melts and retreats into oceans. But in provincial flood control operations, day-to-day, seasonal, and queer, there are probably too many paleochannels in the basin to prepare for avulsion. Assuming Manitobans are not all going to build their homes, farms, and businesses on stilts or mounds, what's a flood control engineer to do?

For that matter, what good does it do that an ethnographer has the knowledge that LiDAR provides? As ethnographer, I find this question easier to answer if I shift my focus from Phase 1, the river's impulse to tear away and engineers attempts to keep it in its channel, and toward the symbolic power implicit in Phase 2, scouting, and Phase 3, flowing otherwise. This ethnography of geo-culture is, after all, an attempt to scout out some basic, practical understandings of engineers and geoscientists in order to find a way of flowing otherwise.[21]

The empirical image of Assiniboine's plurilinear avulsion meshwork lends geophysical gravitas to this book's experimental jumps and juxtapositions. And the knowledge resonates well with Doreen Massey's (2005, 140–1) theoretical approach that winds through this book – the idea that geophysical reality of a place is actually an event, an "event of place," a "coming together of the previously unrelated, a constellation of processes, rather than a thing." Massey argues that we have a responsibility to face up to the fact that even unique places with deep collective meaning may have no intrinsic geophysical coherence. The hyphen in geo-culture holds open this existential gap. In Winnipeg, the glaciers, Lake Agassiz, and the Outburst Quartet lead to the emergence of the Red River and, in its most recent Assiniboine avulsion, the confluence at The Forks. And every spring, the ice breaks up and rivers rise again, independently or together, in a minor reenactment of the original Outburst Quartet, now with new variations. River-city negotiations are ongoing.

III. ICE/WATER AND EARTHWORKS: INSIDE CLIMATE CRISES

How do humans embed cities into earth's river-meshed surface? This book creates a framework, geo-culture, to explore this question in an ice-prone prairie city and province. It imports geoscience and flood control engineering into urban culture, juxtaposing empirical knowledge and the sphere of unintended agencies. By this means, the ethnography characterizes the significance of expert intention as a species-specific mode of orienting collective action while, also, highlighting the agencies of water's elemental bodies. Humans did not intend to trigger climate crises, yet here we are. The contribution of empirical knowledge to survival of all life on earth is crucial. But it is as much about defining the shape of uncertainty as it is about pushing uncertainty away. Identifying the nuances of what we don't know, the how and why, the accompanying senses and feelings, may often be more important than barging ahead with what we (think we) know.

The Outburst Quartet, the geophysical metaphor I use to organize this work, dramatizes the epochal struggle between glacial ice and glacial meltwater. As visual and calculative technologies of geoscience advance, so will the geo-cultural frontiers of flood control. As a symbol, the Quartet slices through ethnographic deep time; as a geoscience image, it's a moving target hitched forever to the ghost of Lake Agassiz (although the lake's name could change, the science seems solid enough). I have taken great liberties in paring down the Quartet for ethnographic use. By collapsing the sequential time spans in which meltwaters break through their ice barriers, it's as if the earth performs a four-movement symphony in the minutes it takes to comprehend the lines on a graphic.

As data continue to roll in, the Outbursts' plurilinear pathways into maritime currents will be better resolved. Indeed, LiDAR already allows Teller, Leverington, and Mann (2002) to produce refinements of the four iconic tentacles. And, too, LiDAR enables Teller and Yang's (2015) correction of some of Upham's 1895 interpretation of Lake Agassiz strandlines, interpretations that had been accepted for a century or more. Even as technologies offer scientists an ever-keener sense of exactitude, knowledge production proceeds on a continent in flux. For example, in a process called isostatic rebound, Manitoba's lakes may be tilting a little less steeply toward the Arctic than they did in the millennia following the outburst of ancestral Lake Agassiz. As polar ice melts, the landforms upon which ice had been pressing renegotiate the balance between buoyancy and gravity that has kept them floating at their postglacial level. This, too, must be folded into the machinery of calculation. For it surely has come to the attention of fisherfolk.[22]

Hydrology, hydraulics, habitat, ethnographic context: all are subject to dynamic change in temporal registers both familiar and strange to our technical traditions. And all have consequences in the sphere of unintended agencies. There is no way for data collection and writing to keep up with real-time riverine action. In the realistic best of possible worlds, an ever-complex dance between received and emerging data-stories will continue to baffle, enthrall, and make work for generations of erratic boulder–strewn, diatom-rich, and pelican-plentiful floodplain dwellers to come.

A. If I Could Write Like a River

If I could write like a river, I, the Assiniboine, would send this proposal to the humanimals:

> *I remember my past lives, cool ... my deep time potential x 8 revealed with LiDAR ... my on-again-off-again relationship with Lake Manitoba revealed by microscopic motile diatom bodies moving toward or away from salt, hey ... I can dig this plurilinear approach to our situation ... it might just be a way to change our expectation ... why not tune my potential peaks and freaks to this deep time visualization? ... Rather than passively, anxiously waiting for the next disaster, me, ... together, we can mimic the crossing of historical and geological time ... Bring it on. Where are the engineers? Let's open up my flow potential, multiply*

the channel x 8 ... so when I'm feeling flashy, I can swell into the meshwork, breathe water into the alluvial fan ... and when the ground is dry and the lake shrinks, I can release back where you need me ... destructive power, transformed, eh? Can we protect the soil on top to replenish but not destroy the pastures, habitats, and crops? ... you've got sensors making data all over my basin ... here, there, everywhere ... tracking my flow, feeling my pulse ... And you've got enough savvy to think hydro-geo-eco-logically ... Now, I know you're going to talk politics, and money ... but look where that's got us ... Imagine instead an avulsion-ready, flash-welcoming set of earthworks with flappable openings that let me come and go, come and go ... while also, always, letting enough of me flow on down to the river city to meet the Red just south of Number-Feet James ... as planned.

IV. WATER GOVERNANCE

I want to share some thoughts about how this book's ideas might be used to reframe and reorganize practical discussions about flood control. I've opened up technical black-boxes in the dynamic trio of geoscience, engineering, and law, translating the inner workings, all while questioning the taken-for-granted and matter-of-fact. What has been revealed should seem recognizable, familiar yet strange, maybe even provocative. I offer a new set of terms, some invented, others borrowed, for remaking the world in this singular and crucial aquatic arena.[23] The terms offer key starting points for public discussions of how expert knowledge and practice could and should be used to inform more egalitarian decision-making, which should in turn lead to different ways of thinking, being, and becoming with rivers. The terms and the questions they generate come from fieldwork conversations, document analysis, and reading transdisciplinary scholarship. They await others who might animate them when the right moments and milieus emerge.

A. Questioning Floodways and Folkways

In the intersecting floodplains of the Assiniboine and Red Rivers, life is suspended in the technozone. Purpose rides in and out of the sphere of unintended agencies, the dimension of matter and meaning within which cities and rivers enact common impulses and within which events often unfold unpredictably. Experts mobilize measurements, models, and codes of geoscience, engineering, and law to operate the control system in the technozone's core. They run the show, but they don't do it alone. Inhabitants are their collaborators, accepting and contesting rules and guidelines (insofar as they are aware of them), attending to and acting on official, mass-mediated communications, especially in emergencies.

The ethnographic concept of blended habitus comes in here: habits and predispositions of individuals scale up to the habitus of communities within the river city; intergenerationally, river-city habitus changes with major shifts in infrastructural history (e.g., before and after the 1950 flood). Experts study deeply enough to become adept in the sociotechnical habitus, the urban floodways of their professions; for the most part, they tend to also be inhabitants whose personal experiences and memories of past floods enliven regional folkways. There is a blend of give and take; floodways are part of folkways. Together, experts and inhabitants contribute to the way that the core system inside the technozone is reproduced and transformed.

So, for example, to what extent might these four mundane technical practices be renegotiated? 1) Number-Feet James is the focal point that organizes the communication interface between forecasters' interpretation of data streams and inhabitants, who are spread out and have to interpret what forecasts mean for them. Like sandbagging, the focal point at Number-Feet James is an important part of Winnipegger folkways (and floodways). What about adding new focal points to the system? With some public education, might this intervention help to inform people who live and work outside the scope of Number-Feet James? Might additional focal points, presented as a hydrologically and culturally informed set, raise a more personal kind of awareness among city inhabitants about the trade-offs entailed in each cubic foot per second of water being kept out of the city, i.e., "their benefit = others' risk?" 2) The organizing techno-metaphor of system balance: is it fair to balance the confluence at The Forks by putting the lakeside dweller at risk? or to put Lake St Martin dwellers at risk to help those farming the shores of Lake Manitoba (a colonial idea still in play)? What might these variously situated infrastructural outsiders think about alternative modes of achieving system balance? How might Indigenous legal scholars and scientists participate in the critical reconceptualization and reengineering of system balance in the technozone? 3) The problem of wind: winds affect water bodies erratically, so it makes engineering sense to drop wind out of calculations for determining state responsibility for harm. But what about prevailing winds? "Prevailing ≠ Erratic." Why not reconsider prevailing winds when setting allowable storage capacity (levels) for Lake Manitoba, for instance? 4) The complications of overland flooding: Manitobans build roads and railway embankments above the prairie, perched high enough to provide safe escape routes or rest stops in floods. But these can also obstruct water that would otherwise flow *away* from households, farms, and small towns, worsening flooding in complex and variable ways. Beyond the calculations that elucidate overland flow dynamics, how, I wonder, do responsibilities for these contradictory impacts get sorted out?[24] While globally, debates about urban flooding tend to revolve around the problems of drainage, zoning, and building codes, these other mundane technicalities should also be discussed.

B. Decolonizing Flood Control

"What do these bodies of water want me to know?"
                Alannah, in Leon and Nadeau's "Moving with Water" (2017, 120)

It isn't the river setting the course of spatial injustice. It's the requirements of monumental modernity. System logic, hierarchical decision-making, immovable infrastructures, and settler protocols combine to produce a hovering technozone setting-in-wait for flood extremes. Inside the technozone, the system works to protect urban inhabitants (in part) by harming-then-helping infrastructural outsiders. This formula of "protected + harmed-then-helped" seems arbitrary, the luck of the draw, where people happen to wind up. And, to some extent, it may be. But it wasn't (isn't?) always arbitrary.

Funnelling excess waters from the intersecting floodplains across an unmarked internal frontier into territories governed by insistent Indigenous sovereignties was an integral part of the province's beginning.[25] Monumental modernity was built on top of colonial earthworks. Sedimenting the harm (irrevocably?) into the system, governmental alliances worked to erase plural legal cultures and river management practices of first peoples, then proceeded to claim common space and common codes. Quietly, persistently, even politely, the Doctrine of Discovery rematerialized the prairie, diminishing sovereignties of all but settlers (Lightfoot 2020). Legally muted, the internal frontier intensifies jurisdictional ambiguities that make it hard for those on the far side of inside, like Lake St Martin communities, to get spatial justice. This is not a recurring "natural disaster." This is not the river's fault. This is a co-optation of riverine logistical power for once nefarious, now impossibly untenable, ends. System-wide decolonization begins by acknowledging the continuing reliance on diverting excess floodwaters on to northern Indigenous communities. Decolonization becomes reality by recognizing, remembering, and restoring communities, ecological habitats, and livelihoods to versions of a future co-created by riverine first peoples who will meaningfully share the power to reengineer contemporary flood-prone conditions. But no matter which loop of the past people wish to re-enter in search of a better footing, restoration has to look forward *and* back.

In river cities, like coastal cities, historical and geological timespaces are crossing now. The avulsion-prone Assiniboine River, whose basin was artificially expanded into the northern lake habitats, sent communiques in 2011 and 2014. Climate change is upon us. River extremes – experienced by humans as excess in floods and withdrawal in droughts – will become more intense, unpredictable, and avulsion-prone. As the Archive and Anthropocene Dialogue makes clear (Appendix I), unwinding colonial water engineering will be key to the great twenty-first-century reorientations required. The Indigenous people whose ancestors were harmed by colonial water engineering have to be invited, with financial and educational resources, into central cross-cultural decision-making roles at the forefront of engineering and law.[26] Decolonizing flood control is essential because it is right and because it makes river-sense.

River cities need protection by engineering systems, but these systems will need to be more flexible. Won't they? Might the larger structures, machines, and decision-making protocols be scaled down and supplemented with others so that willing inhabitant-experts might dance more adeptly with trickster-rivers? Here's a mitigation/adaptation question inhabitants might ask engineers: is there some possibility of inventing mobile structures strong enough to temporarily protect land from floodwaters yet small enough to be distributed more widely and improvisationally?[27]

C. Engineers and the Art of Suspense

Desperately needed and yet quite unprepared, flood control systems around the world are ripe for reinvention even as their enveloping technozones must be revamped to achieve spatial justice. As significant geo-cultural formations that impact the planet, river cities can un-build their infrastructural monuments to modernity and un-settle outmoded land/river divides. The predictable has become improbable, the improbable, essential.

Engineers can be key instigators and mediators in this quest. They could expand the scope of political engineering from its routine operations in backstage enclaves punctuated by emergency performances. Political engineering in the province could expand into unconventional spaces in which expert-inhabitant inter-community and inter-nation collaborations become part of the seasonal round. Engineers could bring together infrastructural outsiders – from Indigenous and settler communities – for wide-ranging talk about assumptions and trade-offs. Together, they could interpretively sift through possibilities with each other and with engineers and political representatives. They might imagine what Manitoba might be like if more of the early settlers – or their more contemporary industrialist and urban planner counterparts – had been inspired by Indigenous lifeways to create hybrid engineering forms attuned to the impulsive agencies of riverine being and becoming. Bring in the artists, the playwrights, the poets, the scholars, the activists, the scientists, the film-makers and letter-writers, the methodologists, the journalists and geovisualizers, the wise people from all walks of life. Gather them from river cities all over the world to imagine flourishing technozones of ecological abundance well back from the tipping points in geoscience models of climate change. Open up the data streams, the designs, the legal tactics of experts to the unexpected, obscured, and unexpressed, and banish forever the cruelties of riverine colonialism. Think outward from rivers flowing across deep time to today – loop back to the Outburst Quartet when erratic boulders hitchhiked and tumbled and the Assiniboine met the vast meltwaters of the Laurentian Ice Sheet to sculpt a delta that became rich agricultural farmland – loop forward into Assiniboine's avulsing possibilities, those grounded visual metaphors that mark the meandering shapes of uncertainty. Yes, it seems I have strayed yet again from my practical intent. But, hey, geo-culture is a stretch. Word.

APPENDIX I

Archive and Anthropocene Dialogue

Cities are deeply affected by and causally implicated in intersections of historical and geological spacetime. When geological processes interrupt human historical processes in and near cities, it can be disastrous (e.g., storms and floods, droughts and fires, earthquakes and tsunamis). But when histories interrupt geology, well, that is a phenomenon that merits a new geological epoch. Enter the Anthropocene Working Group (AWG) of geoscientists.

In the journal *Communications Earth and Environment*, Syvitski et al. (2020) offer a comprehensive review of geological evidence at the regional scale that supports "the stratigraphic case in favor of marking off a planetary-scale Anthropocene time interval at epoch rank, one that would end the Holocene Epoch at ~1950 CE."[1] The Anthropocene Working Group, which includes a number of the article's authors, aims to prove to the International Commission on Stratigraphy, the ultimate authorities on Earth's Geological Time Scale, that this potential new time unit is worthy of being formalized.[2] The AWG and colleagues publish widely. Syvitski et al. (2020) gather quantitative data on a key set of earth surface parameters that collectively form a convincing global signal on the planet's surface, a new geological event and layer of spacetime. The "human footprint" on river systems organized in time intervals is the one parameter among many I recruit for this global-to-regional dialogue.

The general outline of Winnipeg's river engineering history lends local substance to the 1950 cut off. Recall, the Great Flood of 1950 triggered the construction of the modern flood control system in Manitoba. This historical break, a shift within the infrastructural habitus of settler land culture, cuts neatly across regional and planetary scales, deepening the groove of stratigraphic evidence and creating an occasion for dialogue. In this section, I put a collection of key fragments to suggest the ways that (post)colonial law, engineering, and geoscience layered into Winnipeg's diversely inhabited terrain, initiating and installing a recodification of river/land spatial relations.[3] I put this bare-bones sample, a pastiche of territorial precedents, in tandem with the informal interval data that lead to the proposed 1950 Anthropocene cut in Syvitski et al.'s (2020) review. These events happen in a compressed slice of time, geologically speaking. The fragments of this Winnipeg-to-world dialogue punctuate (post)colonial processes, during which, for example, the forests, once spanning out from the great confluence left by the Outburst Quartet at ice age end – itself an event governed by continental scale chaos – give way to a modern settler city with its lovely eighty-acre Assiniboine Park. And all along, the floods keep pulsing.

I tell this story of territorialization backward, with our current predicament and advances toward human rights at the top and deep time at the bottom. This narrative order mimics the settler geology narrated on the Wall Through Time even as it questions the characteristic presumptions of progress (see chapter 1). The story speaks to the imperative of understanding Anthropocene precedents in all their geo-cultural specificities.

Proposed Anthropocene epoch (1950–present): History intersects with geology here, where the AWG proposes to make the cut in the layers of time. Accelerated burning of hydrocarbon fuels and accompanying warming, extinctions, pesticides, fertilizers, and nuclear fallout are some of the more well-known quantitatively measurable traces. These signals are accompanied by the post-1950 "upturn in the global spread of technological knowledge."[4] As technological knowledge spread, so did the hallmarks of river engineering systems, including the construction of dams, reservoirs, and diversions and hardening of levees, with the result that "only 10.5% of large rivers in Europe and 18.7% in North America can be considered as free-flowing rivers."[5]

Law accompanies and has the potential to influence every phase and structure of the post-1950 flood control system. Major components and milestones include: **2018:** Canada signs on to the 2007 United Nations Declaration on the Rights of Indigenous Peoples (UNDRIP), including respect for territorial integrity.[6] **2015:** Canada's Truth and Reconciliation Commission (TRC) calls for the adoption and implementation of UNDRIP as the framework for reconciliation.[7] **2014:** *City's twelfth major flood* came in the summer, not the spring, along the Assiniboine River. Shoal Lake 40 organized activism in tandem with the launch of the Canadian Museum for Human Rights in Winnipeg to raise awareness about suffering caused by Winnipeg's water aqueduct.[8] The body of missing fifteen-year-old Tina Fontaine (Sagkeeng First Nation) was found in the Red River near Alexander Docks.[9] The start of Wa Ni Ska Tan, an alliance of hydro-impacted communities.[10] **2012:** Indigenous scholar Myrle Ballard's doctoral dissertation on the 2011 flood in Lake St Martin and Anishinaabe knowledge systems. **2011:** *The city's eleventh major flood* came in the summer, with the highest water levels and flows in modern history shifting focus to the Assiniboine.[11] **2009:** Flood control system successful in protecting Winnipeg from its tenth major flood.[12] **2007:** Shoal Lake 40 walked 200 km to Winnipeg to protest their reserves' imposed isolation.[13] **2005:** Red River Floodway expansion financed. **1997:** Spring "Flood of the Century," *the city's ninth major flood*, affirmed the wisdom and vulnerability of the floodway. **1982:** Canada's Constitution (section 35) uses the term Aboriginal to refer to First Nation, Métis, and Inuit, although it is not necessarily "embraced by all groups."[14] **1972:** Shellmouth Dam and Reservoir on Assiniboine River, Saskatchewan-Manitoba border, completed. **1970:** Portage Diversion, Assiniboine > Lake Manitoba, completed. **1969:** Signing of the interprovincial Master Agreement on Apportionment, which focuses on sharing sufficient quantities of river water – the main concern of drought-prone Alberta and Saskatchewan – rather than holding back floods – Manitoba's main concern – and, like the 1930 acts, obscures Indigenous sovereignty. **1968:** Red River Floodway completed. **1961:** Manitoba Hydro, which now has fifteen generating stations between Lake Winnipeg and Hudson Bay, founded.[15]

> *Industrial interval (1850–1950 CE):* The interval that "captures the change in human-nature interactions"[16] in which "[m]any large rivers were engineered, with levees, dams and water diversion schemes."[17]

1950: "The Great Flood," *the city's eighth major spring flood*, catastrophic turning point in the history of flood control.[18] **1948:** *Seventh major* *spring flood* in Winnipeg led to flood control planning discussions.[19] **1937:** First sewage system to collect and treat sewage before releasing it with storm runoff directly into the rivers.[20] **1930:** Natural Resources Transfer Acts gave Prairie provinces jurisdictions over water, obscuring Indigenous sovereignty. **Mid-1920s:** Winnipeg officials investigated urban flood protection, "perhaps prompted by J.G. Sullivan's 1922 article in the *Engineering News-Record* suggesting that drainage in the lower Red River Valley would likely worsen the conditions in the city."[21] **1919–21:** Sullivan Drainage Commission created; growing appreciation of the international bioregion created by surface water; Canadian and Manitoban officials overcoming fear of disagreements with their American counterparts as a result of jurisdictional ambiguity. The commission's 1921 report recommended the construction of double dikes on main channels as a means of flood protection. **1916:** *Sixth major spring flood*, during World War I, was followed by "first stirrings of talk about flood prevention."[22] The Greater Winnipeg Water District built an aqueduct to bring fresh water from Shoal Lake 40 First Nation to the city, flooding the Indigenous community and turning it into an island.[23] **1912:** Manitoba's provincial boundaries extended north to Hudson Bay.[24] **1904:** Typhoid epidemic due to reliance on contaminated river water.[25] Dynamite used to release an ice jam on the Assiniboine River in Portage, but Red River floodwaters did not rise above banks in Winnipeg. **1897:** *Fifth major spring flood* in Winnipeg.[26] **1895:** Mandated involvement of engineers and surveyors in rural surface water governance.[27] **1890s:** Canal into Fairford River constructed to provide flood relief to settlers in Lake Manitoba sent excess into Anishinaabe territory.[28] **1888–1967:** Selkirk Ice Movement Record of ice jam events.[29] **1887:** Upham and his assistant do fieldwork in what was once the Lake Agassiz lakebed.[30] **1885:** The Canadian Pacific Railway (CPR) arrived in Winnipeg. **1882:** *Fourth major spring flood* in the settler city of Winnipeg deflated its first real estate boom."[31] **1880s:** Métis prairie communities suffered with the buffalo's collapse and state efforts to suppress Indigenous mobility and clear land for settlers.[32] **1879:** After surveying inhabitants about 1826, 1852, and 1861 floods, engineer Sir Sandford Fleming recommends building the CPR bridge over the Red River closer to Selkirk, but instead, Winnipeg entrepreneurs won the day.[33] **1876:** Legal gender oppression encoded in Canada's Indian Act of 1876 refers to First Nations but not Métis or Inuit.[34] **1874–75:** Date of an official survey map by Dominion Lands Office in Ottawa is a new "visual systematization of land."[35] The authoritative rendering of early city planning shows the downtown grid as well as long lines of plats extending out from riverbanks and across intersecting floodplains (minus the 1872 Indigenous encampment).[36] Measured with Scale of Chains, proposed property lines impose a tri-colour habitat code of areas labelled "under cultivation," "prairie and meadowland," and "woodland" on the flattened landscape. Signatures of the surveyor general, inspector of surveys, and deputy surveyors, together with a stamp by the Canada Department of Public Works, testify to the survey map's legal power. And thus, in becoming a nation, Canada reinforces the imported settlement traditions of the British and French. Like the scraped lines of a glacier moving over rock faces in the continental interior, the pens of the surveyor general and his technical team, indifferent to what and who precede them, inscribe their version of human power on to the earth's surface. **1873:** Winnipeg founded. **1872:** Dominion Lands Act brings steady influx of settlers to take homesteads, mainly Ontario British.[37] **1872–74:** British and United States North American Boundary Commissions hired Métis scouts to guide surveyors and photographers along the piece of their prairie homeland that would become the forty-ninth parallel. The international boundary separating US and Canadian citizens was accompanied by rigid legal distinctions separating Indians and non-Indians.[38] **1872:** Date of a sketch-map, hand-drawn two years after Manitoba was officially recognized as a province by Canada. An encampment of Indigenous people, indicated by three tipis beside a grove of trees and a footpath connecting the tipis to the Assiniboine and the Red near The Forks.[39] **1871:** Indigenous chiefs and Crown representatives negotiated Treaties 1 and 2 at Stone Fort on the Red River in August 1871.[40] Métis people, who had neighbour and kin ties with Cree, Assiniboine, Ojibwa, and Dakota, sought a place at the table.[41] **1870:** Manitoba (name from Assiniboine word meaning "water of the prairie") became part of the Canadian confederation with the Manitoba Act, at the same time recognizing the Métis as a distinct "Aboriginal" people.[42] **Before 1870:** "Overland travelers favoured well-used routes along the natural levees of the Red and Assiniboine Rivers. Farther from the riverbanks, early trails followed ridges and bypassed low areas. Efficient travel usually meant taking the driest – rather than the most direct – line between two places. Though swamps and mudholes too frequently made these routes impassable, they became 'shared resources of the community.'"[43] **1869:** Ottawa ordered the first map survey of proto-Winnipeg – the same year the state would escalate discrimination against Indigenous women. It lives on, embedded as a template in the core of the contemporary street map. Collectively, these colonial map surveys of early settlement show that "the map as an object serves simultaneously as a tool for communicating power (through reference to implicit understandings of space) and an arena for contesting power."[44] **1861:** *Third major post-settlement spring flood*, Red River basin.[45] **1855:** This is the beginning of what Zeller (2000, 87) finds to be the "third, consolidative phase" in which the lessons of entrepreneurial geological imperialism were "applied to broaden Canadian political as well as scientific horizons." **1852:** *second major post-settlement spring flood*, Red River basin.[46]

> *Pre-industrial (1670–1850 CE):* An informal interval that "represents the fundamental transformation from pre-industrial to full industrial energy use" during which "Earth had no discernible climate trend" (Syvitski et al. 2020, 3). In other words, the Anthropocene is set in motion, but the effects of anthropogenic activities have not yet produced a measurable signal.

What's Napoleon got to do with it? In a quirk of history, the settler city was made possible by Napoleon, a human-erratic who inadvertently sets the Manitoban version of the Anthropocene in motion. Here are clues from the layers of documented events: **1850:** Pre-confederate Canadian legislation first establishes a definition of Indian "status" that is bound to patrilineal descent.[47] **1830:** Beginning of "second, organizational phase" in which colonial geologists were inducted "into the social circles of a dynamic, expansive, increasingly entrepreneurial geological imperialism."[48] **1826:** *First major post-settlement spring flood*, Red River basin, still the largest known in history.[49] **1815:** The first phase in which geological thought entwines with political developments in British North America begins, bringing "the colonies into Europe's geological sights."[50] **1812:** First settlers arrived from Scotland through Hudson Bay to settle near what they name "The Forks," with the intention of working the valley's rich black, loamy soil. **1808:** Shut out from European markets, high piles of peltries filled the Hudson Bay Company's (HBC) warehouses with no prospective sales in sight. Turning HBC misfortune to his advantage, Lord Selkirk bought up shares and married an HBC board member's sister. He was rewarded with the concession called Assiniboia that stretched from Hudson Bay to the Assiniboine River and included most of what today is Manitoba and the northern US.[51] **1807:** Lord Selkirk, who wanted to get land in the Red River Valley to support the emigration of poor Scottish farmers, realized that the only feasible way to colonize the valley was to ally himself with one of the two great and rivalrous fur-trading companies, the Montreal-based Northwest Company and HBC. But both companies opposed conceding land to settle farmers.[52] **1806:** Napoleon had been enforcing his "continental policy," a trade war that excluded British goods from European markets. Foreign buyers supplying the Germans, French, and Russians with beaver hats stopped showing up in London.[53] **1776:** The Red River Valley spring flood, part of oral tradition, was not a disaster for Indigenous people because they "did not attempt to live year-round at river's edge."[54] **1670:** HBC, a commercial monopoly and the world's largest landowner, was blessed by Charles II, King of England, Scotland, France, and Ireland.[55] Neither king nor HBC had any sense of the watershed's true scale.[56] HBC assumed the rights to dominate the Hudson Bay drainage basin and the First Nations and Inuit peoples therein, many of whom suffered and died from the influenza and smallpox viruses imported with the immigrants.[57]

Indigenous-precolonial (time immemorial–1670)
[proposed ethnographic informal interval]
The AWG review stops time with the HBC. This is logical. Before the settlers, Indigenous people lived sustainably in their riverine homelands. For the Anthropos to be relevant without being retrograde, the Anthropocene has to include recognition and respect for the light-touch infrastructural habitus that settlers forcibly rejected.

1370 CE: A fragment arises from the official record to suggest the depth of the problem: A moccasin footprint, unearthed from the gumbo under the site of the Canadian Museum for Human Rights in 2008, offers a fraught place-holder for Indigenous versions of human–river relations sustained for thousands of years. The moccasin footprint has been cast in bronze and embedded in a ground floor exhibit. Archeologists estimate that the impression was created by an Indigenous person at The Forks, a seasonal meeting place of First Nations, about 750 years ago (1370 CE).

And with this uncanny footprint I bring the Archive and Anthropocene dialogue to an unsettled end. The historical traces of struggles and of progress, of water and of earth, retrieved and layered into the ethnographic present, gesture, I hope, toward other ways of being and becoming human, urban, and planetary.[58]

APPENDIX II

Technozones

Flood control is a form of provincial territory-making that relies on transforming riverine spaces through engineering. The resulting transformation can be called a *technozone* according to Andrew Barry's (2006) definition: a regulated and infrastructured force field that assumes dominion over spatial expanses by establishing common standards. Its technical practices, procedures, and operational forms distinguish what happens there from what happens elsewhere.

As a landscaped sphere of influence, the technozone is a force and medium of urbanization that casts asunder much of ice/water's ecological and social complexities. Technozones create havoc where they impose themselves on data-sparse spheres of unintentional agencies. The state apportions spaces of hazard and care by creating technozones. But technozones do not dominate in a comprehensive or bounded way. They materialize and regulate through earth-moving and cladding engineering practices, geoscience data, public-private financing, rulemaking, and debate. The more efficient the technozone may be within its scales and spaces of purpose, the more havoc it may create among externalized realms of beings and becoming (from human infrastructural outsiders to mollusks and microbes within).[1]

Transboundary flows in the Red River watershed enroll the United States in Manitoban water governance, affecting the nature of efficiency and havoc and stretching the always uncertain limits of/in the technozones. Consider three examples of international transboundary governance that I've mentioned along the way: 1) the International Red River Watershed Board (see note 1.21)[2]; 2) the Red's origin in the tentacular deep time eruptions of Lake Agassiz linking many of the cities, rivers, and railways between Hudson Bay and the Gulf of

Mexico today (see chapter 2 and Kane 2022); 3) the provincial network of mechanical river flow sensors upon which MIT and Manitoba Hydro depend, supported by a collaboration between Water Survey of Canada and the United States Geological Survey. The Assiniboine River, which is subject to all three international examples, is also governed by the 1969 Master Agreement on Apportionment signed by Alberta, Saskatchewan, and Manitoba (see note 3.8).

Indeed, I find the phenomenon of "the province" to be an illusory yet pragmatic jurisdictional reality with a settler character all its own. Within the version of Manitoban space sketched in figure AII.1, the north gives freedom up for south, edges for centres, micro- and meso- for macro, wildlands for farms and cities.[3] This is what I figured out ...

Fieldwork begins; ethnographer discovers what "everyone around here knows": The legal ideal of the province is a space of care that hypothetically extends to all within its boundaries. The abstract ideal only lives through documents, maps, and performances.[4] I knew this in general. But during fieldwork, I came to appreciate the extent to which territorial *partiality* is not an insufficiency but, rather, the core organizing feature of the functioning province. I also came to fully realize that the province establishes dominion over riverine space in *two* aquatic modes – MIT's flood control infrastructure *protects* the city from rivers, and Manitoba Hydro *extracts* power from rivers.

This much was clear from the beginning: Inhabiting the edge of some of the planet's coldest wildlands, Winnipeggers must contend with the confounding power of ice for much of the year. Frigid temperatures make energy for home heating as central to survival as flood control in this uniquely northern configuration of the water-energy nexus. Relying on monumental modern engineering is elemental. Without it, there would be no way for the diverse population of more than 700,000 to take everyday life for granted either in deep winter or in spring flood.

The confusion that led to a small cartographic experiment: I did not understand how these two provincial systems, with entirely different functions, both pertaining to river flow, did not get in each other's way. So I mapped the monumental infrastructural components of each and drew lines between each point. Two distinct land/water spaces emerged: the flood control system regulated spaces defined by the west-to-east flow of the Assiniboine from the Saskatchewan-Manitoba border to Winnipeg and the south-north Red River Flood-

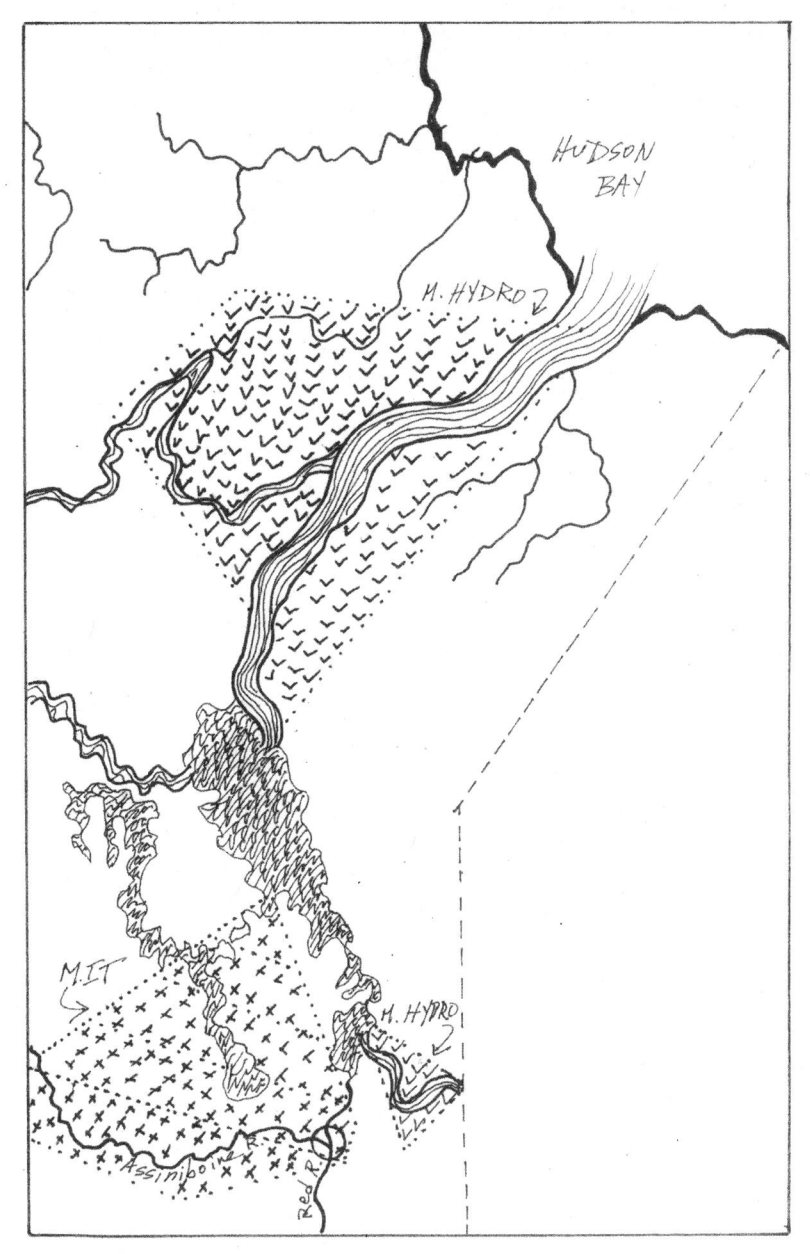

AII.1 Intersecting technozones. Each technozone operates in *the name* of the whole province, but neither MIT's nor Manitoba Hydro's technozones extend across *the space* of the whole province. Nevertheless, they operate infrastructural force fields according to distinct technical standards and practices that have unbounded, multi-scalar effects and affects.

way; the much larger Manitoba Hydro string of dams spanned almost the entire province from the southeast corner through Lake Winnipeg and the Nelson River system to Hudson Bay (with power transmission lines going in the opposite direction, north to south). The former was quite *small* when juxtaposed with the latter (see figure AII.1).

Why it matters ethnographically: According to Barry, a technozone cannot be mapped because its intricate influences are, practically speaking, unlimited (cf. Barry 2006). But mapping can be useful, I think, if the result is not misinterpreted as referencing the whole. In any case, this cartographic exercise showed me that the technozones benefitted the city by occupying distinct and far-flung parts of the greater provincial terrain. It enabled me to see what everyone around here already knows. But the image of territorial fragmentation that emerged from the map did not just fill in a missing piece of reality: it provoked a key insight:

The illusion of provincial-scale dominion is unsupported by the size of the geophysical footprint. Together, MIT and Manitoba Hydro produce the partial fiction that the wet-dry spaces within Manitoba's borders constitute, or merge with, the political entity that is the province. Territorialization, in other words, just needs *enough* terra to sustain the operational effectiveness of provincial organizations responsible for water management. Its power is as much symbolic as material. For those within its grasp, but external to its function, the occupied terra can be terrorizing.[5]

APPENDIX III

Press Conference Transcript

This is a rough draft of the press conference held on 4 July 2014.[1]
The provincial government held a press conference following the morning's declaration of a state of emergency. The *Winnipeg Free Press* posted the videorecording under the byline of Bruce Owen and Larry Kusch, reporters who specialize in covering flood events. "Province to Breach Hoop and Holler Dike. Flooding Prompts Manitoba to Declare State of Emergency."

4 July 2014, posted at 11:35 am and modified at 10:33 pm. Premier Greg Selinger had asked Prime Minister Stephen Harper for emergency military assistance.

[Note: The camera focuses on the officials seated in front. Except for the top of their heads, the reporters below the stage cannot be seen. Time markers, camera shifts, and gestures noted in brackets.]

Premier Greg Selinger reads introduction:

> Good evening. I'm joined by the Assistant Deputy Minister for Water Control and Infrastructure, Doug MacMahon; his Executive Director for Hydrological Forecasting, Steve Topping; and the Assistant Deputy Minister responsible for Emergency Measures, Lee Spencer.
> Since declaring a state of emergency this morning, the flood forecast has worsened. We are now looking at a higher crest, and the water is moving more quickly than anticipated. Our forecasters are now expecting the Assiniboine to crest in Brandon tomor-

row and in Portage la Prairie on Tuesday. Preparations are underway for a controlled breach at the Hoop and Holler bend to reduce the high possibility of an uncontrolled breach. We have updated the Canadian Forces with these new projections and the need for more resources, more quickly. In the last hour, I have spoken to officials in the affected municipalities. Our priority remains insuring the safety of Manitobans. And doing everything we can to [1:00] protect homes and businesses. We will continue to keep Manitobans informed daily of this unfolding natural disaster. I will now turn it over to Doug MacMahon to provide more detail on the updated Flood Forecast.

[The camera does not move from premier to engineer.]

Engineer MacMahon delivers the Flood Forecast and the actions that will be taken in more specific terms. [Note the underlined phrases used to compose the textual fragment highlighting uncertainty in chapter 7 (167).] He speaks extemporaneously, without a text.

Thank you Mr Premier. Yeah, late this afternoon we did receive updated forecast information that did give us <u>indications that we were going to be looking at flows coming into the Portage Reservoir</u> at 54,000 cfs. As is our practice, we do prepare for the upper decile forecast, and that's what we plan to do. As a result of this change in forecast, this morning our flow rates were between 48 and 52,000. This is what we now consider to be the threshold for us to be able to start preparing the Hoop and Holler cut for a controlled breach so that we don't have uncontrolled breaches in the system. <u>We anticipate that equipment is going to be [2:00] arriving at that location in the very near future. And work will commence probably tomorrow morning</u> to ready the location for the controlled breach site. <u>That's not to say we are going to action it. We have to define how the flows come into the site.</u> [Camera shifts to MacMahon and so does Selinger's gaze.] But we do have to prep the site and make sure that the downstream properties are protected from potential flows in the neighbourhood of 4 to 5,000 cfs. This is very similar in nature to what happened in 2011. We have inundations maps that provide us with the potential inundation areas, and we'll be able to push those through the consequence manager so that we can assist people in identifying what properties are at risk. <u>We anticipate that the Hoop and Holler controlled breach location will need to be ready, at the ready, on Monday or Tuesday</u> in order to control some of the flows. [3:00] Works are also continuing on increasing the capacity of the Portage Diversion to make sure that it can handle substantial flows. In 2011, we had it ready for 34,000 cfs. <u>That's our plan, and we're still on track to achieve that</u>. The Assiniboine dikes east of Portage la Prairie are also being inspected, and hotspots are being actioned. These will continue until the peak of Tuesday or Wednesday next week. That's pretty much the update for today. This is late-breaking information, and our forecasters are updating the forecast as the information unfolds. <u>This is a summer rainfall event, and the forecasting is a lot more dynamic than winter forecasts</u>. But we are updating the forecast continually. I'll turn it over to Steve Topping to provide the detailed forecasting information. [Camera moves to put Topping in the frame, MacMahon turns gaze toward Topping and then into space.] [4:00]

Engineer Topping reads the specific forecast in technical terms. He looks up occasionally:

Yes, today we notified Brandon that they could receive a forecast peak of anywhere between 37 and 38,000 cfs tomorrow. Water levels will be slightly higher than 2011 levels. For instance, in 2011, Brandon saw 36,730 cfs. They do have adequate flood protection in Brandon to deal with this type of flood due to the works that have been constructed since 2011. The Assiniboine River at Portage Reservoir is forecasted to peak at 52 to 54,000 cfs on approximately July 8th to 9th. And this is assuming ideal weather. Water levels will be similar or a bit higher than in 2011 for areas downstream of Portage la Prairie. [5:00] The expected crest will be fairly sharp [Topping's hand gesture emphasizes the idea of a sharp crest and looks more directly as he says:] in that it will probably be at 50,000 cfs for a day to two days in duration. That concludes mine [turns back to the premier].

Premier: Lee ...

Assistant Deputy Minister Lee Spencer speaks with no notes, looking straight out at camera and audience:

Thank you, Mr Premier. Obviously there's been a change in our forecast, and there's a change in the approach we are taking. We have taken the opportunity to talk to all the municipalities that are affected by [mike noise] Hoop and Holler, and we'll be distributing the inundation maps this evening to those municipal governments to help them begin notifying those people in the inundation zone of the fact that they will need to make some flood protection. With the army arriving and with this additional workload beyond just preparing for 18,000 on the river, we expect to begin provincial works to help those people in the inundation zone [6:00] protect their property as early as tomorrow. As the army begins to arrive, we will be fanning out with resources to help those identified properties. In priority, working from the river downstream to prepare for the potential Hoop and Holler cut. This will be something done in cooperation with the municipal government and using the military as our backbone, I suppose, to make sure we are able to meet the very tight timelines necessary to protect the property in the matter of three to four days. [Turns back to premier.]

Premier: Questions?

[6:38]

Questioner 1: Will as many properties need to be protected as in 2011 when not a lot of water hit many of them?

[Premier scratches his head and looks at Spencer.]

Spencer: The inundation zone planned for 5,000 cfs is larger than what was impacted in 2011. We are going to plan for 5,000 cfs in the inundation zone again. The people that were warned in 2011 will be the same people that will be warned this year.

Questioner 2 (woman's voice): ... There were about 150 homes in 2011, will it be the same?

Spencer: I don't know the details on that. But it will be the same.

Premier [writes something down, looks up with tongue in cheek, brow furrows]:

He made an important point [both hands raised, extends index finger of one hand]. We prepared for 5,000 cfs in '11 but we didn't have to go that far. So it's the same level of preparation [moves hands back toward his body and then forward, gesturing repetition]. We are not sure how far, how close we'll get to that.

Woman Questioner 2: 5,000 cfs. Can you put that in average laymen's terms? Explain for us [laugh], how much water is that?

Premier: Steve ... [everyone on stage turns to look at Topping, but the camera holds on the premier.]

Topping: That's a tough question.

Audience member: It's about half of what the river was carrying until early this week, [7:34] until early this week.

MacMahon: That's a good analogy. [Nodding his head in agreement.]

Questioner 3: As of this morning, you said 300 homes will need to be protected. Has that number changed?

Spencer: No, that's the same number. That's the area downstream and the area along the Assiniboine River. There is an additional 150 to 200 properties and farm infrastructure that will need to be protected on the Hoop and Holler inundation map. We're just working now with the municipalities to identify those properties. [8:00]

Woman Questioner 2: What would happen if you didn't make this cut? How many homes? What kind of devastation?

Premier: The risk of an uncontrolled breach – and then you lose control of what's going to happen – that's a very key point. And once you lose control, you don't know the magnitude of how much damage will be done and people and lives are at risk so ... The original Hoop and Holler cut was made in 2011 in order to be able to manage extremely high flows of water. The same rationale applied today. It's all about minimizing and controlling the way the water goes and to manage it and control it and to know ... An uncontrolled breach could occur anywhere along the diking system, and then you've lost complete control of what's going to happen to people.

Questioner 3: Do we have the sandbags to do all this?

Premier: Yes, well, we had two million in reserve, half a million starting up today, and add another million to that right now.

Questioner 3: What happens in summer, folks are not at home, so ...

Premier: Military people are going to be available, and so if that property needs to be protected, they can come in and do that. In cases of emergency.

Questioner 3: Do you have any details on how many troops and from where right now?

Spencer: The vanguard of the initial ... will come from CFC Shiloh. The size of it is not confirmed yet, and it will be banding now based on the initial work that we are asking them to do. The initial deployment is coming from Shiloh. I expect that we will see troops from other parts of the western region. That's something the military will decide based on their own readiness levels and who's ready to go across their command structure.

Questioner (not clear which): And the number?

Spencer: That's expanding. I don't know the number yet. I can't answer that.

Woman Questioner 2: Do you know how many extra hands ...

Spencer: Well, ideally, we have somewhere in the neighbourhood of 350 properties to protect and three to five working days. That's a challenge for anyone. But with volunteers, with the property-owners themselves, and with the military, we are going to do our best to meet those targets. I've spoken to the commander of Joint Task Force West this afternoon and warned him of the additional workload and asked him to begin to deploy more than he had initially thought were necessary.

Questioner 3: That's 200 identified this morning plus 150 ...

Spencer: Approximately 150 in the Hoop and Holler inundation zone.

Questioner (not sure which one): What about Winnipeg? The water is going toward Winnipeg, so what is going to happen to Winnipeg?

Premier [shifts body into position]: Once the Hoop and Holler is open, if it reaches the LaSalle River, there's the potential for water to enter Winnipeg on the LaSalle River just north of the floodway. I have phoned the mayor, and I have informed him of that, and these officials have been informed at the equivalent level at the city. But we've informed everybody and ... [turning to look to Spencer and then MacMahon and Topping] Do you have any sense of what would be the effect on property in Winnipeg if that occurred? [camera stays on the premier]

Topping: 18,000 cfs on the Assiniboine River the city of Winnipeg will have to prepare for, and generally that involves gravity storm water sewer systems [camera shifts to Topping] – blocking them off and putting emergency pump systems in place.

Woman Questioner 2: Can you give people a sense of what that would look like?

Topping: [shrug]

Woman Questioner 2: Along the side of the river …

Topping: The outfalls of the storm water sewer systems would be either gated or blocked off, and emergency pumps would be made available to pump water over primary dikes along the Assiniboine River.

Woman Questioner 2: I just want to understand how the water goes. Eventually where? And it impacts Winnipeg, why?

Premier: [camera holds on Topping, MacMahon, and the premier] If it reached the LaSalle River it could impact Winnipeg through the LaSalle River just north of the Floodway. Steve [premier turns to Topping], do you have any sense of what that would do to the 18,000 [rolls hands in gesture of approximation]. How much more would that add? [12:00]

Topping: Well, actually, it'll enter the Elm Creek ditch and get into the LaSalle River system [tracing key interconnections with hands on table], go into St Norbert, and with the floodway we would be able to quite easily manage that additional flow.

French-Speaking Questioner: Premier, might you translate this into French? [my rough translation]

Premier: [my rough translation] In French, the situation is very serious this evening. We have the possibility of too much water in our system and in preparation for this, it is necessary to prepare Hoop and Holler to be opened to protect the community of Portage le Prairie and other communities east of Portage la Prairie until Winnipeg.

Questioner: [unidentified] What's the priority … in Hoop and Holler? But does the same priority exist in the Portage la Prairie area on the dikes? Are we going to see troops in both places at once?

Premier turns to Spencer [camera holds on the other three].

Spencer: Yes, we have to get ready for both. 18,000 cfs down the channel and Hoop and Holler. We will be deploying troops to both tasks as they arrive.

Questioner: [unidentified] This might sound like an odd softball question, but this is déjà vu all over again, and I'm just wondering if there's any personal thoughts on how we're going to go through this all over again.

Premier: Well, no one expected to have this kind of an incident three years out since the last one. It's obviously tough on people. We feel for the people who are in the way of the water; we want to make sure they are protected. So our hearts go out to them. But more importantly, our efforts go out to protect them. So everyone's gearing up with the same enthusiasm and dedication as they did last time. And I guess we have the advantage of some experience now. We have a clear idea where the water is going to go. The reeves and the volunteers know the task ahead of them. We have better inundation maps now. But clearly it's a strain on everybody. But clearly people are in a sense ready.

[14:00]

Questioner: [unidentified] What are you going to do with Lake Manitoba? It's just slightly above flood stage right now.

Premier: Again ...

Questioner: [unidentified] continues: People are rebuilding ...

Premier: Again, there has been a lot of rebuilding going on. Some communities are better protected, got permanent dikes, some properties ... We had a program for individual property-owners to protect themselves better. We changed our standards for any rebuilds. They have to build up to eight [feet] twenty-two [inches] I believe [turns head toward MacMahon] is the standard now. Steve? Now the big challenge on any [lifts hand in warning] lake, Lake Manitoba, Lake Winnipeg, are wind events. They can add a crisis very rapidly if they hit at the wrong time and the wrong place.

Questioner: [unidentified] What about the rain this weekend? What kind of role does this play in your forecast? [camera stays on the premier, who is drinking water.]

Topping: As I said, our forecast is best on ideal weather. We are forecasting due to the high Humidex,[2] the possibility of thunderstorms. These thunderstorms tend to be localized, which right now, they're forecasting southeast Manitoba to get thunderstorms, so that would not be at risk. So it's all dependent on where those storms are centred. If one were to occur over Brandon, it could create problems.

[Unseen organizer says to someone: "I want to make sure we get to the phone line."]

Premier: OK, let's go to the phone line [technical guy comes up to the table].

Caller: Brandon is reporting that there could be a crest on the Assiniboine that could last for weeks. Can you talk about that at all?

Premier: [nods in agreement] Steve?

Topping: We are forecasting a second crest of approximately 32,000 cfs which would be in the timeframe of July 17th and 18th, and water levels will generally stay high in the Brandon area. After the first crest, [shrug] generally in the order of about 30,000 cfs for a couple of weeks, two to three weeks.

Any other questions?

Phone Questioner: Hi. Can you hear me? [Yes.] It's Gene Holtrey, *Enterprise News*, calling. Regarding Lake Manitoba and the St Martin channel that was built in 2011. Engineers reported back then, recommended not only building that channel but also building a separate smaller channel around the Fairford Dam. And my question is, why was that not built around the Fairford Dam to begin with? Because I think if that were there right now, I think there would be a lot less stress on people living along Lake Manitoba.

Premier: I'm going to ask our PDM responsible for water control works, Doug MacMahon, to respond to that.

[MacMahon and premier in camera frame]

MacMahon: Yeah, that was one of the scenarios we looked at back in 2011. It's a very difficult solution to implement. It's First Nation property. There's bedrock. It's very technologically challenging. The second challenge that you have there if you increase the flow out of Lake Manitoba, you'd have to increase the flow out of St Martin. And the system balance wouldn't be there. For us to action that, to respond to this event is not practical in the time frame we are talking about.

Premier: You should know that we are planning a second outlet from Lake Manitoba. And Doug, can you update us where we're at on that?

MacMahon: We're in the throes of doing a conceptual design [premier closes his eyes and smiles] for the enhanced outlet out of Lake Manitoba. We were prepared to go to public input in June, but because of the events that have unfolded over the last week or so, we've delayed that until the fall. The engineering for that is whittling down the alternatives that we could look at as realistic alternatives and the next step for us to get public input onto it. It is a very major initiative and a very expensive initiative for our future water control structure, infrastructure.

Phone Questioner: What do you have to say to the people who live along Lake Manitoba?

[Videotape ends here.][3] [18:19]

Notes

CHAPTER ONE

1 Every body or thing – a river, a boulder, a city, a person – is always intent on being and becoming. This idea guides recent scholarly efforts to recognize and understand the powers and agencies of material things and bodies. Seventeenth-century philosopher Spinoza (1992) calls it *conatus*. In my reading, it is Jane Bennett (2010, 2–5) who carries this word into the realm of urban water infrastructure when one sunny day, her perceptual field shifts. Upon glancing down, she is struck by the vitality of a "contingent tableau" composed of glove, oak pollen mat, dead rat, bottle cap, and stick caught in the grate over a storm drain (5). In his book *Spatial Justice*, legal theorist Philippopoulos-Mihalopoulos (2015, 7–8) intervenes theoretically to argue that while the vitality of conatus connects all bodies, it is also what allows bodies to withdraw. Spinoza's concept of conatus resonates with Indigenous ideas about water bodies as relational beings with spirit. For entryways, see feminist Indigenous scholars Yazzie and Baldy's (2018, 2) introduction to radical relationality, spatial justice, and a politics in which "World view is water view, a view *from* the river not a view *of* the river," and Cameron, Leeuw, and Desbiens's (2014) critical essay on the use of "Indigeneity and ontology" in cultural geography. Gordillo (2021) briefly brings together Indigenous perceptions and Spinoza to reorient Elden's (2010) political theory of territory. Conatus-like ideas are active well beyond scholarly arenas, of course. For example, recent Indigenous struggles to decolonize river governance have led to the enactment of legislation that accords personhood status to rivers in Bangladesh, Colombia, Ecuador, and New Zealand.

Ideas about being and becoming in human–water relationships wind through a number of recent edited volumes whose authors inform my scholarship, including, for example: *Arts of Living on a Damaged Planet* (Tsing et al. 2017), *A World of Many Worlds* (de la Cadena and Blaser, eds 2018), *Rivers of the Anthropocene* (Kelly, Scarpino, Berry, Syvitski, Meybeck, eds 2018), *Territory beyond Terra* (Peters, Steinberg, Stratford, eds 2018), *Infrastructure, Environment, and Life in the Anthropocene* (Hetherington, ed. 2019), *Hydrohumanities* (De Wolff, Faletti, and López-Calvo, eds 2022).

2 For recent books on water; infrastructure; and cities, the state, and nation-building, see, e.g., *Concrete and Clay* (Gandy 2002); *The Meaning of Water* (Strang 2004); *In the Nature of Cities* (Heynen, Kaika, and Swyngedouw, eds 2006); *Impossible Engineering* (Mukerji 2009); *Material Powers* (Bennett and Joyce, eds 2010); *Fluid Pasts* (Edgeworth 2011); *Disrupted Cities* (Graham, ed. 2010); *Confluence* (Pritchard 2011); *Reigning the River* (Rademacher 2011); *Where Rivers Meet the Sea* (Kane 2012); *Empire of Water* (Soll 2013); *Beyond the Big Ditch* (Carse 2014); *Roads* (Harvey and Knox 2015); *Endangered City* (Zeiderman 2016); *Public Infrastructures / Infrastructural Publics* (Collier, Mizes, and von Schnitzler 2016); *Hydraulic City* (Anand 2017); *A City on a Lake* (Vitz 2018); *A Future History of Water* (Ballestero 2019); *The Invention of Rivers* (Da Cunha 2019); *Engineering Vulnerability* (Vaughn 2022).

3 Bourdieu (1985, 79) writes that habitus is "the *intentionless invention of regulated improvisation* [italics in original]." I extend Bourdieu's anthropocentric theory to include rivers. As rivers move through the geophysical formations that city dwellers relate to most durably (e.g., primary dikes in central Winnipeg), they improvise as they co-create the shape and meaning of regulated material reality. Rivers negotiate the different historical episodes of habitus along with differently situated human groups (see also note 1.49). When flowing through Winnipeg, rivers, I suggest, have encountered three shifts in infrastructural habitus that resonate globally, from the dramatic break between plural Indigenous ways of being and becoming with riverine "amphibious nature," to the embanked and drained earthworks of settler "land culture" (cf. Hooimeijer 2014 on the Netherlands), to the technology-based systems of "monumental modernity" (cf. Nixon 2011, 150–74, on India). See also Klaver (2017, 20) on the "naturally hypernatural authenticity" of the Dutch landscape.

4 In geo-culture, space and time cannot easily be separated. I switch between the terms spacetime and timespace, depending on the empha-

sis in any particular setting. My analysis intersects with intellectual traditions associated with merging time and space post-Einstein through Bakhtin's concept of chronotope (see note 1.48 and chapter 7).
5 This description draws loosely from Upham (1895, 108) and Teller et al. (2002). See chapter 2 for a fuller version of this story.
6 Along with the Bering Land Bridge theory, scientific naming practices are in play. For a historical overview of scientific naming, from Linnaeus's naming of mammals (seemingly emerging from the youth's not totally logical focus on breasts) to the geologists coining of the Anthropocene, see Schneiderman 2015.
7 Dispossession or mass killing is often followed by symbolic rematerializations that valorize what was lost, as in the transformation of bison from real herds into Manitoban logos. For a Singapore River case in which people are dispossessed of traditional lifeways by the modernizing state, only to have their traditions rematerialized as art and heritage, see Kane (2018a, 142–3).
8 Geoscientists tend to think of water bodies as inert matter, lacking vitality or sentience, yet their models have the peculiar capacity to transcend their methodological and philosophical underpinnings. The science upon which the Outburst Quartet is based seems almost shamanic in its metaphorical capacity to bring a non-specialist like me in contact with the extraordinary drama of continental-scale deep time events. But it is nevertheless important to take the geopolitical interests of globalized geoscience knowledge networks into account (Cruikshank 2005; Cameron 2017). As Yusoff (2019, 8) argues, geologists of today inherit a compromised nineteenth-century language that continues to silently carry the work of resource extraction, racism, and ongoing settler colonialism into the world (see chapter 2).

The troubling racial politics associated with naming Lake Agassiz is a case in point: In light of current Indigenous and anti-racist politics in North America, the nineteenth-century naming of Lake Agassiz troubles the origin story presented here, as does the reproduction of the Agassiz name in contemporary geoscience and culture more broadly. A scientifically constructed ghost, Lake Agassiz has been a formative character living in the Manitoban geological imaginary since the nineteenth century. Has its vast, empty, yet persistently measured aquatic shape lent geophysical form to settler projections of *terra nullius*? Reading the name in another co-occurring register, the Agassiz name sutures colonial crimes against Indigenous people in Canada with slavery and anti-black racism in the United States (see, e.g., McKittrick 2021, 154–60; cf. Wynter, n.d.).

The name of Lake Agassiz was given by Upham (1895), the geologist whose fieldwork measuring its abandoned terrain was inspired by the ice age theory successfully promulgated by the Swiss natural historian and Harvard professor (see chapter 2). The Agassiz name was also given to landmarks and institutions well beyond Manitoba. However, when Tamara Lanier, a descendant of the enslaved subjects of photographs that Agassiz commissioned and Harvard held in their collection, revealed the situation to the public, the contentious fact that Agassiz used his scientific work and professional standing in a way that supported white supremacy could no longer be ignored. The public began calling for removal of his name in various places. (For example, see https://www.insidehighered.com/news/2019/03/21/lawsuit-against-harvard-focuses-actions-19th-century-biologist; Stevens, Heidi, *Column: Harriet Tubman Will Take the Place of Racist Scientist's Name on Lakeview's Agassiz Elementary School*, chicagotribune.com). For earlier scholarly discussions, see *Mismeasure of Man* (1981) in which paleontologist Stephen Jay Gould reviews racist nineteenth-century theories in zoology and in the new field of anthropology, including Agassiz, and historian Irmscher's discussion of racism in his Agassiz biography (231–3).

Although I follow current naming usage, I do wonder if ongoing Indigenous rights and anti-racist efforts to change public culture may raise questions for Manitobans. Should the ancient glacial lake continue to bear the Agassiz name? Would it even be possible or healthy to disentangle the name of Lake Agassiz from Manitoban cultural history or glacial science? If so, what kind of collective effort might devise a new name for a ghost lake never encountered in its living extent by humans? Perhaps "Turtle Island," a translation of the name for the North American continent in some Indigenous oral histories, might serve as a departure point (https://www.thecanadianencyclopedia.ca/en/article/turtle-island)? If pursued, these questions could become a geo-cultural extension of mapping and naming efforts that restore Indigenous people and place names in North America; e.g., see Thomas and Paynter (2010) for Manitoba. See also the Pan-Arctic Inuit Atlas at http://www.paninuittrails.org/index.html?module=module.about. See also *Anijaarniq*, the cinematic geovisualization of Inuit wayfinding based on Claudio Aporta's dissertation fieldwork: https://www.anijaarniq.com.

9 Indigenous in appearance, they are within walking distance from the small Dakota Tipi settlement upstream, on the other side of the reservoir.

Notes to pages 7–13

10 Following Liboiron's (2021, 3, note 10) anti-racist naming practice, I resist relying on an unmarked identity that "re-centres settlers and whiteness as an unexceptional norm."
11 See video of pelicans fishing in the Red River at Lockport, Manitoba: *White Pelicans of Lockport, Manitoba – June 27, 2017*. https://www.youtube.com/watch?v=htOtREu-UJo.
12 For background, see: Manitoba Infrastructure. https://www.gov.mb.ca/mit/wms/pd/index.html.
13 This paragraph is inspired by Bennett's (2010, 4–8) opening to *Vibrant Matter*. While happening to notice the contingent tableau assembled in a grate over a storm drain (see note 1.1), she is suddenly struck with the idea that this stuff "commanded attention in its own right, as existents in excess of their association with human meanings, habits or projects." She calls this "thing-power."
14 For forensic and many other applications of diatom knowledge, see Smol and Stoermer (2010).
15 I developed this approach while participating in two research networks of scientists, social scientists, humanities scholars, and artists. Now an edited volume (Kelly et al. 2018), *Rivers of the Anthropocene* grew out of a 2014 conference at the Arts and Humanities Institute of IUPUI in Indianapolis. Events of the Ice Law Project, convened regularly between 2014 and 2019 by the Centre for Border Research, investigated the potential for a legal framework to govern Indigenous homelands and global shipping routes in frozen Arctic maritime regions. See archival materials: https://icelawproject.weebly.com.
16 Inspired by Mol's (2002) innovative ethnographic practice.
17 Aradau and van Munster (2011) study how insurance companies use aesthetic technologies in anticipating catastrophic risk. They define aesthetics as action upon the senses.
18 There are two theories of the cause of this troublesome section of the asphalt road: 1) a neighbourhood theory, commonly put forth for trouble spots throughout the city, involves an underground spring (shared both by a neighbour and Pauline Greenhill); and 2) mechanical engineer Kendall Thiessen's hydrogeological theory that a curving mass of clay slipping off the bank into the river is causing the instability (see chapter 3). In 2017, another stabilization project was completed; accompanying the project description is background on the structures I am describing here. See https://www.winnipeg.ca/publicworks/construction/majorProjects/lyndaleDrive.stm#tab-background.

For ethnography about the techno-imaginary of underground water sources and infrastructures, see Kane (2012, 23–40, 61–81, 93–115); Anand (2017); Ballestero (2019).

19 I come to see the functional beauty of the primary dike from University of Winnipeg geographer Bill Rannie on the introductory city infrastructure tour he gave to me and geographer Nora Casson (14 August 2014).

20 In Winnipeg, ice affects everyday life in myriad ways, although it is not an essential feature of survival as it is for people farther north. In the Arctic, for example, Inuit who live north of Manitoba in Nunavut and the Canadian Arctic Archipelago rely on frozen rivers and seas for navigating while hunting, fishing, and visiting (Aporta 2009; Aporta et al. 2018).

21 For international governance issues, see the International Red River Watershed Board of the International Joint Commission (https://ijc.org/en/rrb).

22 In dynamic landscapes, sustaining (the appearance of) accuracy is not only a problem of having to transpose scales but of methods used to do so. Ferrari, Pasqual, and Bagnato (2019, 21) write: "representations of territory are bound both to the capabilities of the systems that generate its data and to the intent of the observer." The extreme precision of today's tools, like GPS and LiDAR, make it difficult, they find, to disguise the fact that the glacial terrain underlying political borders in the Alps cannot be truly fixed. See chapter 7 for how LiDAR disrupts Manitoba's conventional geological imaginary.

23 MIT is now called MTI, Manitoba Transportation and Infrastructure. In the book, I keep to the name used in my fieldwork year of 2014.

24 Alf Warkentin, meteorologist and retired MIT forecaster (interview, 20 October 2014).

25 Ibid.

26 Gill draws on Said's (1994, xii–xiii) analysis of land as object of imperial battles. On mapping and colonialism outside Manitoba and Canada see e.g., Jacob (2006, 269–360, cited in Mirzoeff 2011, 58); Wood 2010, 34; Stransbjerg (2012, 826); Dodds and Nuttall (2016, 97, 161–2); Da Cunha 2019.

27 Historian of science, Zeller (2006, 382) analyzes events that set Manitoba up for its colonial encounter with European mapmaker-scientists. The geographer-explorer Alexander von Humboldt's "'cosmic outlook' [and isotherm mapping method] circulated in Canada to refine territorial expansionists' scientific arguments justifying annexation of Rupert's Land after the monopoly of the Hudson Bay Company's [HBC] expired in 1869." HBC had dominated Rupert's Land since 1670 when the Doc-

trine of Discovery gave King Charles II the ideological back-up to dispossess first peoples and granted to the HBC the entire drainage basin (an initiatory gesture with only a vague geographic referent). For in-depth analysis of implementing UNDRIP and unwinding Canadian colonial history, see Borrows 2019.

28 Michel Hogue (2015, 1–2) argues that the photographs of the Métis engineers ("sappers") with their mounds not only documented the progress of British and United States boundary commissions but "established the lines and measurements that would serve as the basis for subsequent land and railroad surveys." Elden (2019) notes the arduous nature of translating the forty-ninth parallel from someone's abstract choice to a visible line in his essay on boundary-marking. Making meaning out of the spaces between empirically observable points is a theme that recurs in geopolitics, geoscience, and this ethnography.

29 By radically questioning "the grip that the river and by extension, the line separating land from water, had on [his] imagination," Da Cunha (2019, ix) argues that lines on maps grant rivers "the status of permanent residence" on land but make "water in other moments of the hydrological cycle ephemeral, so that rain, mist, snow, humidity, and so on are seen as visitors whereas rivers are granted the status of residents (9)."

30 For example, a hydrology map of Manitoba published not long after the construction of the Floodway illustrates hydrometric data for all the province (Weir 1983, 17). If one looks carefully, it is clear that except for northern rivers measured for hydroelectric power infrastructure, data was only collected in southern Manitoba, which ultimately becomes the flood-protected zone.

31 Steve Topping (with Chris Popper, 2 October 2014; with Darrell Kupchik, 25 November 2014); Alf Warkentin (20 October 2014).

32 Bumsted (1997, 8), I discovered later, had already written about the "geographic confusion" I experienced. He explains that the Red is one of the few rivers on the continent that flows north. The result is some geographic confusion, since the so-called upper reaches of the river are to the south and the lower reaches to the north.

33 The Northern Divide (a.k.a. the Laurentian Divide) separates the Red River that flows *north* to Hudson Bay from the Red River that flows *south* to the Gulf of Mexico. Unlike the Great Divide of the Rocky Mountains, which sends river waters *east* to the Atlantic Ocean and *west* to the Pacific Ocean, in North Dakota the Northern Divide is located on fairly level land. https://www.ndstudies.gov/gr4/geology-geography-and-climate/part-2-geography/section-7-continental-divide.

34 Folding this phenomenon into Bourdieu's (1985) general definition of habitus (see note 1.3), I would say my field experience reflects cross-cultural differences in geospatial or geomorphological habitus. Caveats: as a hypothetical aspect of our earthly sensorium, such differences in human–river relationship would be subtly situated, not ever-present, easily decipherable, or useful as a common basis for social categorization.

35 The proposed cross-cultural difference associated with the Northern Divide (see note 1.33) corresponds roughly with an international divide, since Canadians inhabit most of the Arctic-tilting terrain. The difference may rarely be noticed, but it is not neutral. In *North of Empire*, Jody Berland (2009, 50–1) critiques historical and policy representations of north-to-south river flows as "natural" and shows how they contribute to the marginalization of Canada with respect to the US. By implicitly orienting technical knowledge, might such aspects of elemental water–human relationships shape future international negotiations regarding the global hydrosphere?

36 See also Wood (2010) for how the map projections that influence how we occupy and experience space are laden with preconceptions about that space and thus become the centre of controversies.

37 Massey (2005, 239) cites Open University (1997, vol. a, 80) for this fact. (I cannot find the original full citation.)

38 This understanding should disturb the systematic bases upon which local place-based knowledge is collected and processed to create reliable global knowledge (as in biome and anthrome classification systems based in remote sensing of regional differences, e.g., Ellis and Ramankutty 2008). The larger question is, how should we negotiate the tension between systematic collection of local-place-based knowledge and distant or speculative geological and ecological processes in anthropogenic fluvial landscapes? For another kind of natural history, see, e.g., Raffles 2002.

39 We have significantly reordered the world's river systems (Williams et al. 2014; Kelly et al. 2018; Grill et al. 2019). For the Mississippi River, the most famous North American example, see Barry 1997.

40 Humans are differently situated with respect to the Anthropocene and the idea of humans as geological actors. The "here" of Indigenous people in Manitoba has been violently destabilized since colonization, and yet the many thousands of Indigenous people who live in Winnipeg also benefit from flood control that sends excess water to the northern territories. In *The Accident of Being Lost*, Leanne Betasamosake Simpson's (2017a) storytelling powerfully and humorously conveys the contradic-

tory feelings and thoughts of an Indigenous person living in a Canadian city. However, if you are in a city hit by an extreme flood, whether you are a settler, an Indigenous person, or a visitor passing through, you may well appreciate the technical traditions of flood control. This appreciation appears in one of Simpson's stories about experiencing a flood for which inhabitants were not prepared. In this context, she recalls (108–9) that she had met her partner in Winnipeg's Flood of the Century in 1997 and writes: "Winnipeg, you know how to sandbag. There are stations, there are hundreds of volunteers, the bagging of the sand takes place in the same place as the sandbags are needed, and people know how to architect sandbags into amazing triangular walls of steel." (See chapter 3 for more on sandbagging as an important part of blended habitus.)

41 I import the term "potential Anthropocene" from Syvitski et al. (2020, 2), who argue for the Anthropocene as a new "distinction in the stratigraphic record" (see more below and in the "Archive and the Anthropocene," Appendix I). While still awaiting official geoscience recognition, the term has already transformed cultural thinking and triggered considerable debate (e.g., Haraway 2016; Cameron 2017; Whyte 2018; Yusoff 2019). Here, I cautiously use the Anthropocene's "potential" to set the stage for renegotiating the unfolding confluences of historical and geological time.

42 The terms "forks" and "confluence" can refer to the site where rivers join together, a junction. Technically, however, if one thinks with the direction of riverine flow, this central Winnipeg site is a confluence, because the Assiniboine is flowing into the Red, not branching off of it. And yet, geo-culturally, inhabitants experience and navigate the confluence as a place where things happen. And thus, water creates a sense of process, place, and event on the planetary surface.

43 Like bridges that carry travellers above and across the rivers and their ambiguous edges within modern cities, the wall embodies and symbolically reproduces the social order by steering inhabitant mobility (Winner 1977, 127, cited in Kane 2011, 223). For further exploration of ambiguous geographies, see Peters et al. 2018.

44 The wall disrupts natural processes: as trees die from standing for months in floodwaters, as happened in 2011, they send out adventitious roots, readying future versions of themselves to begin again, in a new space. Trees on the upland also send rhizomes and seedlings downward. In this way, the old river bottom reasserts itself as forest. I learn this from Kristin Tuchscherer, city naturalist, at the Assiniboine Park

Restoration project. Despite her optimism, she says that ultimately, "Humans can't replicate nature" (site visit, 9 September 2014). If such concrete-walled engineering is done in a concentrated and comprehensive way, as in Singapore's renowned freshwater hydrohub, which cuts almost all the island's rivers off from the sea, the result is what I call "enclave ecology" (Kane 2017), a defining character of metropolitan nature.

45 The wall was built by volunteers from the International Union of Bricklayers and Allied Craftsmen, Manitoba chapter. See: https://www.winnipegarchitecture.ca/wall-through-time and https://www.tyndallstone.com. For geological origins of Lake Winnipeg's Tyndall Limestone, see Russell (2004, 20, 16–29). Thanks to Pauline Greenhill who pointed out beautiful examples of Tyndall Limestone gracing Winnipeg's architecture as we took Assiniboine River neighbourhood walks.

46 For geoscientists' imagination of hypothetical post-Anthropocene observers, see Williams et al. (2014, 62). For a geographer's critique of "a mythic Anthropos as geologic world-maker/destroyer of worlds," see Yusoff (2016, 3).

47 The Forks, as a "mixed use" socio-economic hub, was not developed until the 1990s. https://www.theforks.com/about/the-forks. Before that, the rivers had more room to flood without threatening the city. A second civic monument at The Forks also promotes the embodiment of flood history. http://www.mhs.mb.ca/docs/sites/redriverflood levels.shtml.

Globally, floods are an ancient and persistent theme in art (Withington 2013). However, the creation of flood milieus as art installations for public immersion may be a new climate crisis genre. See Dan Roosegaarde's virtual flood light installation named *Waterlichte* in Toronto. https://www.studioroosegaarde.net/project/waterlicht.

48 The wall draws me to use tools from literary theory: Following Jameson (1981, 17–102), I approach the wall as a set of collective texts from a "political unconscious," texts that should be read in a series of broader interpretive horizons. The most salient broader horizon in this wall-reading is the Anthropocene. The wall = a thing = a text. Upon the wall's surface lies a conventional text, a historical narrative, that expands, even as it pins down, the wall's meaning. Also related here is Mary Louise Pratt's (2017, G170) analysis of the Anthropocene concept as "what narrative theorist Mikhail Bakhtin called a *chronotope*, a particular configuration of time and space that generates stories through

which a society can examine itself." In chapter 7, I analyze an emergency press conference as a chronotope in greater depth.

49 Quoted terms from Hooimeijer's (2014) episodes or phases of Dutch building traditions. Note that, in this Manitoban analysis, there is a clear shift in infrastructural habitus of land culture(s) between pre- and post-1950 flood control. From settlement to 1950, the city, like outlying farming communities, protected themselves with dikes and drains. In the most extreme big floods, the city would be overcome, and people had to evacuate. Post-1950, however, the monumental engineering structures protected city inhabitants, creating a sense of safety that materially and emotionally contributed to the modern infrastructural habitus. The wall does not mark the shift but rather presents land culture as a progressive continuum.

As the city developed, infrastructural habitus dramatically shifted for the worse in Indigenous communities as the city increasingly shunts excess floodwaters into the spaces beyond the northernmost edge of the control system. Thus, cross-cultural differences in riverways persist, change, and produce different tensions and language to express and negotiate these tensions. For the case of Lake St Martin, see Ballard (2012) and chapter 5.

In practice, the episodes of urbanized infrastructural habitus don't follow each other neatly in bounded segments of timespace. Rather, episodes exist simultaneously in different material forms, degrees, and patterns of distribution. Nevertheless, historical periodization can, I suggest, help to clarify some of the tensions in (post)colonial "contact zones" (Pratt 1992). There have been distinct infrastructural tensions – not only between settlers and Indigenous nations but also among settlers (between colonial-era fur traders and settler-farmers; contemporary city dwellers and farmers). In her comprehensive history of drainage and governance among farmers in Manitoba's wet prairies, Shannon Stunden Bower (2011, 141) writes: "The history of catastrophic flooding along the Red River surely must be one of Canada's most compelling illustrations of the hubris of European settlement." Moreover, she finds that intermittent inundation in the wet prairie is not a problem that can be solved but rather an environmental reality to be managed (170).

50 Forks Redevelopment 2000; 1950 flood; railway lines, diesel locomotive, steam plant, buildings, bridges, cathedral.

51 1826 flood, flour mill, experimental farms, Métis farmsteads, Selkirk settlers, paddlewheels, and forts, including Stone Fort (a.k.a. Upper Fort Garry).

52 Indigenous villages, camps, and hunting parties of the Assiniboine, Cree, Ojibway, Dakota; Métis traders; La Vérendrye (military-explorer-fur trader) built Fort Rouge 1738–41.
53 Camps of the Assiniboine, Cree, and many others unnamed, inter-tribal meeting between 1550 and 1600; flood and drought between 1000 and 1500; bison; grasslands.
54 Lake Agassiz figured as a preternaturally calm series of wavy layers. There is no evidence of the Outburst Quartet's ice/water violence.
55 Glacial retreat, full glaciation, glacial advance, mammoths.
56 Indigenous scholars and allies who are collaboratively creating new ways to imagine the future include, e.g., Tsing et al. 2017; de la Cadena and Blaser 2018; Whyte 2018; Craft and Regan 2020b; Fujikane 2021; Liboiron 2021. Interestingly, the name of the Assiniboine means "allies" (or stone water people) (Leon and Nadeau, 2017, 120). These collaborations stretch into applied settings as well. For example, the community- and land-based research of the "Decolonizing Water" project based in British Columbia enacts a "two-eyed seeing" approach to water monitoring that "blends Indigenous and Western ways of knowing." (https://decolonizingwater.ca/our-approach/#two-eyed-seeing).
57 The Wall Through Time, as I interpret it here, is one of myriad minor architectural contributions to the cultivation of modern infrastructural habitus – i.e., a human-river orientation that integrates a strong, protective city-focused engineering response. At this moment, however, this strong engineering response is 1) not ready for the climate crisis (an existential fact shared by river cities worldwide); and 2) contradictory to Canada's Indigenous-led commitments to the Truth and Reconciliation Commission (TRC) process and the implementation of the United Nations Declaration on the Rights of Indigenous Peoples (UNDRIP).

But the civic space of The Forks enacts cross-cultural solidarities as well as the contradictions of colonialism. The nooks and crannies of settler geology that I bring to light here subtly undermine but do not override the upland areas of The Forks that highlight Indigenous artists and honour Indigenous cultures more broadly. The Oodena Celebration Circle, for example, works in infrastructural and symbolic counterpoint to the Wall Through Time below it. Along the Circle's edge, the monument for honouring missing and murdered Indigenous women and girls serves as a living memorial space of sorrowful critique and resistance. See https://www.kairoscanada.org/missing-murdered-indigenous-women-girls/monuments-honouring-mmiwg. https://www.theforks.com/uploads/public/files/attractions/oodena_info.pdf.

58 To construct the dialogue, I searched through the data gathered during and after fieldwork, e.g., all the fragments I could find on my few visits to the Manitoba Archives and Winnipeg Library and on all the dated touchstones I could find in the scholarly, popular, and governmental corpus in my bibliography. I created a scaffold of extreme flood events and the dates of major engineering and legal infrastructures, then layered in the telling details of territorialization and resistance I had at hand. In this sense, the contents are somewhat arbitrary. The dialogue could be opened up at any point and taken in multiple directions and readings. That said, the Appendix is meant to gesture toward the kind of comprehensive historical research that may illuminate the unfolding geo-cultural events implicated in the planetary-scale signals of twenty-first-century trouble. It points toward a potential future of grounded geo-cultural analysis of water bodies and human bodies.

59 Following Simpson (2008, 32, cited in Craft 2014, 6–7, note 20).

60 Rannie (city infrastructure tour, 14 August 2014).

61 This topic of conversation didn't come out of nowhere. During a site visit to the Assiniboine Restoration Project (in a remnant of the great forest that is now Assiniboine Park), Kristin Tuchscherer shows me how to read signs of past flooding on the tree trunks, the whitened bark below separated from the darker above. As I pull my gaze around the forest, the line on the bark connects the trees a few feet up from the river bottom; below the line is a subtle but present enveloping mist – the ghost of the flood of 2011. (Interviews and site visit with Tuchscherer, 9 September 2014.) See also note 1.44. (Conversation with dog-walker, Lyndale Drive, 10 September 2014).

62 I'm following Bill Rannie's advice to think like a hydrogeologist and "figure out where the water wants to go" (city infrastructure tour, 14 August 2014).

63 The sheer size and flooding power of the Red in comparison with the Assiniboine has positioned it in the forefront of flood control.

64 Although there was also some flooding within the protected zone. On the city infrastructure tour, Bill Rannie takes us to upscale homes on sites that should have been subject to more restrictive zoning changes (14 August 2014).

65 Thanks to Jason Senyk, policy analyst, MIT, for giving me a copy of this reprinted photograph collection (interview, 10 November 2014). For historical photos of the 1950 flood, see also Bumsted (1997, 45–77), and for a video snapshot, see Pauls 2019.

66 Beverley Peters, born and raised in Brandon, tells me about the anxieties

of rural households when they take in ethnically different strangers from Winnipeg in the 1950 flood (interview and site visit to Omand's Creek, 9 November 2014).

67 For example, Paul Thomas, political studies scholar, recalls the 1950 flood through a pair of uncanny transport images (Lambert 2020): "We couldn't travel on the highway at one point. My grandfather went on the railway track driving ... with one wheel inside the track and one wheel out, bumping along ..." and "We have family photos of my dad paddling a canoe into the living room in our modest storey-and-a-half home." Thanks to Terry Zdan for sharing this article with me. https://www.nationalobserver.com/2020/01/03/news/dikes-ditches-and-dams-manitobas-fight-against-flooding-complex.

68 This discussion draws on Mukerji (2009, 214–15), who defines logistics as the "mobilization of natural forces for collective purposes." She distinguishes between the state's use of strategic and logistic power: "Strategies are efforts to organize human relations while logistics are efforts to organize things ... they bring different types of power to social life." In this way, "[i]nfrastructure changed the political significance of nature."

69 On the 1997 Floodway situation, see Thomson (1997, 21) and the Z-dike story in chapter 3.

70 Bill Rannie (city infrastructure tour, 14 August 2014).

71 For example, see photo archive of the 1950 and 1997 floods on the city's official flood history pages: https://winnipeg.ca/History/flood/flood_scenes/flood_1950.stm and https://winnipeg.ca/History/Flood.

72 See Shrubsole (2001) for earlier discussion of cultural aspects of modern flood control based on analysis of the 1997 flood.

73 With political geographers and writers of "new materialism," I insert the dash in geo-political to emphasize politics of the earth rather than human-centred politics of nation-states, although they do intersect. For a brief historical overview of the dash, see Peters, Steinberg, Stratford's "Introduction" (2018, 5).

74 Cities from the Great Lakes region of North America to Siberia come to mind. That said, climate change is altering flood dynamics everywhere, making various aspects of Winnipeg's expertise in dealing with unexpected riverine action widely relevant. In July 2021 alone, as I write this note, catastrophic flooding events have occurred in urbanized continental interiors of Africa, the Americas, Asia, Europe, and the United Kingdom. Australia was hit in February 2022. For details, see https://floodlist.com. The IPCC (Caretta et al. 2022, 4–72) published their latest assessment of phys-

ical evidence at the global scale. In section 4.4.4 on "Projected Changes in Floods," the scientists write:

> In summary, the assessment of observed trends in the magnitude of runoff, streamflow, and flooding remains challenging, due to the spatial heterogeneity of the signal and to multiple drivers. There is however high confidence that the amount and seasonality of peak flows have changed in snowmelt-driven rivers due to warming. There is also high confidence that land use change, water management and water withdrawals have altered the amount, seasonality, and variability of river discharge, especially in small and human-dominated catchments. https://report.ipcc.ch/ar6wg2/pdf/IPCC_AR6_WGII_Final Draft_Chapter04.pdf.

75 Infrastructural outsider status may intersect with, and be submerged in, the class, race, and gender parameters of dominant majority identity groups, e.g., communities north and west of the Portage Diversion or south of the Floodway. In contrast, Indigenous infrastructural outsiders do not fit into majority paradigms in the same ways, and this changes everything (see chapters 4, 5, and 7).

76 Erasure and the intentional confusion of identity is a critical theme that can be traced widely in the scholarly literature. Putting the discussion of infrastructural outsiders in Judith Butler's (2016, 6) philosophy of gender terms, for example, I am suggesting that the norms governing flood control do not operate to produce infrastructural outsiders per se as "recognizable" persons. Opening his chapter titled "Unimagined Communities: Megadams, Monumental Modernity, and Development Refugees in India," environmental humanities scholar Rob Nixon (2011, 150–74) writes: "If the idea of the modern nation-state is sustained by producing imagined communities, it also involves actively producing unimagined communities." In *Water Wars*, Vandana Shiva (2002, 15), the physicist-turned-environmental activist upon whose work Nixon draws, makes visible the techno-geopolitical actors and methods responsible for dam-building and water scarcity. She recognizes the humanity of the people whose lives are affected, who name themselves "oustees," as a critical part of rejuvenating "ecological democracy." See also Anand's (2017, 8–10, 55–9) analysis of how poor migrants in Mumbai do infrastructural work on their homes to establish their belonging and "hydraulic citizenship," thereby resisting their classification as "outsiders" unworthy of adequate municipal water services.

Ordinary people may become activists as a means to struggle directly against and reverse engineered unrecognizability. In Buenos Aires, I observed neighbourhood activists from different parts of the city with different water infrastructural problems recognize each other and come together to organize politically (Kane 2012, 130–76). Might infrastructural outsiders from Manitoba, like the Porteños of Buenos Aires, overcome the fragmentation characteristic of their relations with the government while sustaining the recognition of the significant differences between settler minorities and plural Indigenous cultures and communities? Highlighting the difficulties of this work, Michelle Daigle (2019) finds that even scholars who may critique erasure of identities may miss or confuse their own complicity. She confronts the often hollow "spectacle of reconciliation" with Indigenous people performed on Canadian university campuses. Decolonizing the domain of flood control could ground reconciliation and re-cognize outsiders in ways such spectacles never will. Pushing these thoughts further into the Black anticolonial terms of Katherine McKittrick (2021, 154–60, citing Wynter n.d.), might the experience of infrastructural outsiders – their "geographic estrangement" – help to unsettle the categorical terms of settler colonialism more widely?

77 These scholarly spheres also overlap with political geography and political geology (Bobbette and Donovan 2019).
78 Interview, Steve Topping and Chris Popper, 2 October 2014.
79 For an example of how the Winnipeg model is being considered for other Canadian cities (with historic footage from 1950 and Floodway construction), see Pauls 2019.
80 For an example of how Singapore's freshwater hydrohub, an enclave ecology, may unintentionally but negatively affect the island's estuaries, see Kane 2017 and note 1.44.
81 See also notes 1.29 and 1.30.
82 See many mid-twentieth-century examples in *Manitoba Atlas* (Weir 1983). In probably the most dramatic twentieth-century infrastructural transformation, Manitoba Hydro (2011), a provincial Crown corporation, diverted the entire Churchill River into generating stations along the Nelson River in the 1970s. In my interview with Dale Hutchison (17 September 2014), a Manitoba Hydro natural resource manager, he explained that at the beginning, the engineers had a pioneering attitude (as in "you did what needed to be done").

The rights of First Nations or the requirement for assessing the environment were "not around" then ... We no longer divert rivers, he said, because now we recognize the rights of First Nations and we are required to do environmental assessments. But that we did it then enables us to add more stations ... now. [Paraphrased from handwritten notes.]

83 Nixon (2011, 162–3) identifies five main strategies, which may be used in combination, to deny the rights of "hydrological zone" inhabitants (I paraphrase): 1) threat of direct violence; 2) rhetorical appeal to self-sacrifice for the greater good; 3) denying hydrological zone inhabitants' rights and diminishing their visibility (by adopting the guise of neutrality, e.g., calling the dispossessed "project affected" in their documents); 4) dismissing inhabitants of a projected submergence zone as culturally inferior – a strategy that "did not end with the waning of direct colonialism" and that links "irrational"/ "lawless" people with their "irrational"/"unregulated rivers"; 5) abrogating the rights of the dispossessed. Nixon is writing about World Bank dam projects in India, but these water governance strategies are not so foreign to Canada. Indeed, the strategies have been globalized.

84 Interview with Myrle Ballard, natural resource scholar (8 November 2014).

85 For scholar-activist-community resistance in Manitoba see "Wa Ni Ska Tan: An Alliance of Hydro-Impacted Communities." http://hydroimpacted.ca/wa-ni-ska-tan-2.

86 The term "sphere of unintended agencies" can work as shorthand in interdisciplinary space. For example, it could refer to Judith Butler's (2016, 18) idea that "there is a vast domain of life not subject to human regulation and decision, and that to imagine otherwise is to reinstall an unacceptable anthropocentrism at the heart of the life sciences."

87 I also used merged terms to switch emphasis as needed, e.g., spacetime or timespace, expert-inhabitant or inhabitant-expert, and I coin the "Outburst Quartet" as a simple way to invoke a complex hydrogeological origin story throughout the narrative.

88 While classic ethnography resides on a representation of the "local" grounded in human-to-human social interaction, this experiment animates the local by narrating hydrogeological knowledge through storytelling. Boundary crossings to wider worlds and worlds within weave through, resonating from backstage. To bring river-city actors alive, myriad other possible water bodies and infrastructures are left out. The limits to the book's complexity and scale are necessary fictions not that different from limits in the design of scientific experiment and interpretation (see examples in chapters 2, p. 53, and 8, p. 184).

89 See Sauchyn et al. (2020) for a literature review of climate change impacts and adaptations in the Canadian prairies, including intersectional social dimensions. See Prairie Climate Centre (2019) for Winnipeg climate atlas: https://climateatlas.ca/sites/default/files/cityreports/Winnipeg-EN.pdf.

90 See note 1.3 on Bourdieu. The concept of blended habitus also builds on Raymond Williams's (1977, 132) concept "structure of feeling." Williams writes:

> We are talking about characteristic elements of impulse, restraint, and tone; specifically affective elements of consciousness and relationships: not feeling against thought, but thought as felt and feeling as thought: practical consciousness of a present kind, in a living and interrelating continuity. We are then defining these elements as a "structure": as a set, with specific internal relations, at once interlocking and in tension. Yet we are also defining social experience which is still *in process*, often indeed not yet recognized as social but taken to be private, idiosyncratic, and even isolating, but which in analysis (though rarely otherwise) has its emergent, connecting, and dominant characteristics, indeed, its specific hierarchies.

I draw on the structure of feeling concept when writing about approaching floods and, more specifically, when writing about inhabitant-expert attunement to lake levels in chapter 5 and the tension between fear and confidence in chapter 7. The book's title, *Just One Rain Away*, conveys this structure of feeling. More broadly, the concept of structure of feeling enters scholarly currents in diverse ways. For example, Yazzie and Baldy (2018, 6) distinguish their specific emphasis on the "cacophony of struggle" from "decolonization as an intellectual and political methodology" that "has become a normalized structure of feeling (Williams 1977) in Indigenous studies and Indigenous political formations." Different structures of feeling, such as those mentioned here, can generate frictions in city life.

91 Tomiak et al. (2019, 2) write: "The mythic separation of the city from its surrounds in settler colonial discourse – which imagines the city and the reserve/reservation as completely disconnected spaces – renders invisible the violence upon which settler city-building relies." I suggest that internal legal frontiers support the violent process by which such spaces appear to be disconnected. The concept could thus be helpful in making them visible.

92 The aqueduct transporting fresh water from Shoal Lake 40 to supply Winnipeg is another case of (post)colonial water body inter-linkage

that disadvantages Indigenous infrastructural outsiders (Tomiak et al. 2019, 2).
93 See notes 1.48 and 7.4 on chronotope.

CHAPTER TWO

1 Before professional outsiders arrived to study the glaciers, Bjornerud (2018, 135) explains in her book about geological time: "Swiss farmers understood that large boulders strewn far down alpine valleys marked the former positions of ice masses." In his biography of Agassiz, Irmscher (2013, 40) writes: "Standing in a landscape to which they didn't belong, these masses were regarded by locals as 'strangers to the soil.' Locals called them *Findlinge* (German for homeless children). So, although Agassiz gets much of the credit for ice age theory, the original, albeit unremarked sources of fundamental scientific knowledge were farmers and hunters who intuited the relationship between erratic boulders and ice fields in the observant course of their everyday lives." (See more of the Agassiz story below.) See also Cruikshank (2005) for an ethnographic analysis of how Indigenous knowledge of glaciers is produced as "local" in colonial-era Yukon Territory.

2 The year of erratic naming seems more certain than the name and profession of the person who named them. Hughes (2019, 125, 129, citing Vivian 2001, 72) cites Ignace Venetz, the naturalist who would give erratics their name and argue that they could only have been deposited by the glaciers that once carved them. However, the *Oxford English Dictionary* documents another 1828 work, a quote by W. Phillips <u>Treat. Geol.</u> (Humble), "The magnitude of the transported rocks is such as to deserve the name of erratic blocks." Rudwick (2009, 105–6) names Prussian Leopold von Busch as the first geologist to try and decipher the puzzling erratics of the Alps and northern Europe, publishing his theory of a (not necessarily biblical) "geological deluge" in 1815. In this essay, Rudwick (2009, 106) publishes an 1820 sketch now in the National Museum of Wales. The artist, Henry De la Beche, depicts a tiny man with an ax next to a giant granite erratic that had travelled about 100 km from Mont Blanc to the Jura Mountains. Even as it astonishes the viewer, the function of the sketch presents important empirical evidence in support of von Busch's theory. That said, Rudwick doesn't actually say that it was von Busch who named the puzzling blocks "erratics." (See note 2.3 for more on erratic art.)

3 Mediated by local knowledge, erratics enter the course of European art and architecture as well as science (Hughes 2019). In the 1830s, not long after philosopher-natural historians began deciphering the relationship between erratics and glacial passage and well before the ice age became accepted scientific theory, Karl Friedrich Schinkel, the architect who would build Berlin's Altes (Old) Museum, was in the Jura sketching erratics when he became inspired to act:

> Schinkel famously "carried," or ordered the carrying of, a seventy-five ton erratic block, the Granitschale (itself cleaved from a 750-ton erratic mass), to the entrance of his Altes Museum and had it carved into a fountain. Too big to get through the portico it remains deposited in front of it, in Berlin's Lustgarten – a testimony to another force, this time of architectural will, carving its way through the urban landscape (Hughes 2019, 128).

Berlin's erratic holds a signifying power that combines glacial and human forces. Schinkel directs the erratic, not simply to be moved from the place the glacier left it but to be forcefully cleaved from an erratic ten times its mass. And too, Schinkel did not simply have the hewn boulder placed in a prominent garden for all to admire its natural magnificence. He had unnamed sculptors turn it into a fountain so gigantic it made miniatures of strollers-by. The fountain's waters still dance on its polished surfaces. Nailed to the erratic fountain, a marker points to the relation with the royal art collection inside the museum, now a UNESCO World Heritage Site: "The block is the subject of Johann Erdmann Hummel's uncanny painting *Die Granitschale im Berliner Lustgarten* (1831), now in the collection of the Alte Nationalgalerie." http://pointurier.org/travel/germany/berlin/museuminsel/altenationalgalerie/slides/IMG_0067.webp. See drawing of the original 750-ton erratic (the larger of the two): https://deacademic.com/pictures/dewiki/71/Gegensteine_auf_den_Rauenschen_Bergen.jpg.

Erratics travel across generations to figure prominently in contemporary conceptual art. They do so in embodied form: for example, Jessica Ramm's Erratic Boulder "from Perthshire is strapped down and made mobile" in a Saatchi Art gallery. (See https://www.saatchiart.com/art/Sculpture-erratic-boulder/721533/2101579/view.) And they leave their bodies behind to travel the imageric world, e.g., as canvas prints produced by a wide variety of artists in different locations. https://fineartamerica.com/shop/canvas+prints/erratic+boulders. In "Traces of Another Time in the Landscape, 2017," Maria Whiteman brings gallery visitors into videoed relation to a human hand, "caressing" an erratic's crevices and curves. Grunewald Gallery, Indiana University Bloomington, 11 October–20 November 2019.

4 From contemporary Earth Art (or Art Povera) to satellite-generated colour graphics in geoscience journals, aesthetic tools are central to the construction and disruption of the humanist values and empirical models dominating institutional art and science spaces. Sketching Bruno Latour's (2013) argument in *Inquiry into Modes of Existence*, Bobbette and Donovan (2019, 7) write:

> aesthetic tools are mechanisms for seeing and acting with places at a distance; of displacing places for us and transporting us to other places. They make the earth sensible as a story of material transformations that can be registered by the human body, while opening the body outwards to the expanses of geological time.

In short, the figure of a human body standing against (or maneuvering) a puzzling erratic boulder is a founding image that opens the everyday to other spacetimes.

5 For example, during the expedition to demarcate the US/Canada border between Ontario and Michigan (1819–27), a British physician, John Bigsby, reported "vast spreads of erratics around the shores of Lake Huron" (Rudwick 2009, 108)."

6 In this chapter, Part II presents Lake Agassiz geology through Upham's empirically based poetic vision. Part III presents some of his findings as a hydrogeological foundation for understanding Manitoban river–city relationships.

7 There is a measure of speculation (a willingness to stray from factual to fictive versions of the unknown) in the construction and interpretation of geological data. As an expert cultural orientation, willingness to speculatively extend knowledge aligns geologists with forecasters, who have no choice but to extrapolate in moments of crisis. Their relative comfort when travelling into the sphere of unintended agencies distinguishes geoscientists from geotechnical engineers whose responsibility for building reliable earthworks like dams and bridges discourage speculation of any sort. (Readers will meet such forecasters and engineers in later chapters.)

8 Alas, other than his one assistant, this second-generation geologist from the United States has no merry team of natural history students and artists to encourage and delight as does Agassiz in the Alps.

9 See note 1.8 on the politics of (re)naming the glacial lake after Agassiz.

10 Irmscher (2013, 39–40) writes that "where Agassiz imagined sheets of ice, both Darwin [following Lyell] saw icebergs drifting in water."

11 Like naturalists collecting fauna in the Amazonian river basin earlier in the nineteenth century, Upham's scientific practice actively participates in an international project of "narrativizing ... geography" (136) that relies on the "mutual indispensability of reason and aesthetics" (Raffles 2002, 131).
12 For critique of the sense of colonial privilege evidenced by Upham and continuing among scholars and scientists today, see Whyte (2018). For contrast, see Liboiron (2021, 19), who strives to "scientific and/or activist work that does not reproduce colonial L/land relations." See also chapter 1 and note 1.76 on colonial strategies of erasure.
13 Compare to the British and United States North American Boundary Commissions that hired Métis scouts in 1872–74 to guide surveyors and photographers along the piece of their prairie homeland that would become the forty-ninth parallel (Hogue 2015).
14 The name of another glacial lake, Ojibway Lake, encodes their dominion. Ojibway Lake merged with Lake Agassiz in later phases of its history (Teller et al. 2002 and references therein).
15 Mentions of Winnipeg include mostly unanalyzed details gleaned from other authors and added at the end of the book. The topics include wells, fresh water, layers of limestone, alluvial clay, sand, and gravel (200, 526, 576–7); rain, snow, and other weather patterns (592–3, 599); historic flood of 1828 (598); agriculture among Ojibway, Dakota, and Scottish immigrants arriving in 1812–16 (610–12); limestone quarries (626).
16 Quote from J.A. Elson (1983, 24), a mid-twentieth-century geologist studying Lake Agassiz.
17 For historical context, see "Archive and Anthropocene Dialogue," Appendix I.
18 In 1914, with Upham back home, the opening of the Panama Canal usurps global trade, and Winnipeg slides into a period of economic decline. The historic lull enables the cut stone and terracotta warehouses built during the boom to survive and to become the thriving art-filled historic Exchange District today. For more on Winnipeg as past and future node of global transport, see below.
19 On the other hand, there are explorers, fur traders, and settlers intent on bending the knowledge practices of nineteenth-century science toward the building of empire. See Zeller 2000; 2006; Grek Martin 2014; Byrne 2016.
20 Compare the ancient Nazca of Peru, who create the Nazca Lines by mimicking the material language of the earth. By painstakingly sorting

white pebbles from dark over enormous distances, their graphics mediate sky and water.
21 The striae and erratic boulders are among his proofs for ice sheet expansion along a north-south gradient (in contrast with the east-west gradient Agassiz observes in Europe – see Irmscher 2013, 95–6).
22 Thanks to geoscientist Kimberly Rogers (email 24 June 2018) for interpretation of Upham's use of these terms.
23 This sketch comes from Russell (2004, 22), who draws from various sources including interviews with contemporary geologists James Teller and Eric Nielson.
24 This is the site of the spillway scene in chapter 1.
25 In other words, Lake Agassiz is a central trope organizing popular and scientific "geographical imaginations" (Gregory 1994, 168–9, cited in Berland 2009, 78).
26 Quotes and definitions of this term in the *Oxford English Dictionary*: 1892 G. De Geer in <u>Proc. Boston Soc. Nat. Hist.</u> XXV. 457. "To get a general view of the warping of land ... I have used the graphic method of Mr. G. K. Gilbert ... and have connected with lines of equal deformation, or as I have called them **isobases**, such points of the limit as were uplifted to the same height."
27 For example, Humboldt's 1817 isotherm map (Hughes 2019).
28 From a geomorphology textbook (Bierman and Montgomery 2014, 448–9):
   Shoreline and delta locations can be combined into maps and the uplift amounts contoured. These isobases or lines of equal uplift, can be used to infer the mass distribution of the now-vanished ice and the magnitude and orientation of the uplift field.
29 See Ferrari et al. (2019, 47–9) on the history and politics of land measurement devices and the politics of cartographic imagination. As part of their larger experimental art + history + science project, they connect sensors picking up and transmitting movements of Alpine ice edges (= international borders) to a drawing machine in an art gallery (174–92). See also Part IV of this chapter on the entwined destinies of melting glaciers and the city.
30 In this respect, scientific knowledge production is not unlike the cosmopolitics of shamanic practice, as I argue in *The Phantom Gringo Boat* (Kane 1994). Both modes of discourse bridge the gap between known and unknown forces and enroll diverse sentient beings.
31 For in-depth analysis of this issue in contemporary global climate change policy, see O'Reilly, Oreskes, and Oppenheimer (2012).

32 This section is a version of Kane 2022.
33 National Ocean Service. Currents; Thermohaline Circulation. https://oceanservice.noaa.gov/education/tutorial_currents/05conveyor1.html.
34 This section is a version of a piece of text from Kane 2022.
35 Website https://centreportcanada.ca/about-centreport. Thanks to Terry Zdan for bringing CentrePort to my attention and to John Spacek, its vice president, for the interview. Terry Zdan interviews, 16 September and 7 November 2014; Spacek interview 21 November 2014.
36 Latitude and longitude lines also expand outward from the pole in the original geoscience article. Neither map goes much south of the Gulf of Mexico. See Haraway (2016, 32) on the "tentacularity" of "life lived along lines," which in her analysis refers not to rivers or railways but rather to living creatures of all kinds.
37 Most tri-national map images for the USMCA, the new NAFTA, signed by the US into law in December 2019, stick to variations of the old NAFTA map. Compare Jane Henrici (2002), who argues that "some of the specificities of space along this [the NAFTA] corridor have come to seem less significant than the speed of contact in between." Current events can disrupt even the most entrenched corridors and specificities in the space of global trade. As I edit this manuscript one last time in March 2022, Canada is stripping Russia from its "most favoured nation" status because of Russia's war on Ukraine. This throws plans for linking Winnipeg and Murmansk via the Arctic Ocean into question.
38 The Dene call the mightiest one Deh Cho in the Slavey language. The name means mighty river, or more precisely, according to Kulchyski (2005, 120): "land of the people of the river, people of the land of the river, river of the people of the land; the meanings circle each other, surround each other, repeat each other, reflect each other. Like the sound of a drum. Like the dance it inspires." The English name is Mackenzie after a Scottish fur trader and explorer. Flowing into Deh Cho is Naechag'ah. "Nahe" means "powerful" in both a material and spiritual sense. Nahendeh refers to the region, territorial government, and voting district.
39 See note 1.1. On stones as "object[s] of wonder," see Klaver (2005).
40 An insight inspired by the Argentinean art collective "Los Erroristas" (a.k.a. Etcetera).
41 Frank Pearson, manager of Mining Division, Winnipeg Supply, described the scene at the pit in "History of the Rock." http://uwinnipeg.ca/rockclimb/history-of-the-rock.html.
42 In April 2022, the Hudson's Bay Company donated its vacant landmark

building to the Southern Chiefs' Organization, representing thirty-four Anishinaabe and Dakota First Nation communities. https://scoinc.mb.ca/a-new-future-wehwehneh-bahgahkinahgohn.

For another Canadian example of an impressive erratic migration marked with a bronze plaque, see sculptor Maura Doyle's *There's a New Boulder in Town*, the result of "a ten-ton moss-covered boulder transported from a spot in eastern Ontario" to the Toronto Sculpture Garden. https://sculpture.org/blogpost/1860252/348457/Between-a-Rock-and-a-Hard-Place-Maura-Doyle-and-the-Glacial-Erratic.

43 This sketch is a composite of details from Irmscher (2013, 38–9, 74–5) and the Mount Auburn Cemetery, http://mountauburn.org/2011/louis-agassiz-1807-1873. Irmscher (2013, 343–4) also includes the comic side of Agassiz's fall from grace – a picture postcard of his statue toppling headfirst off Stanford University's Zoology building in the 1906 San Francisco earthquake. For more erratic art, see notes 2.3 and 2.4.

44 Rocks with stories are "material semiotic actors" in Donna Haraway's terms, but their peculiarities, in Jane Bennett's words (2004, 348, 351), are "not entirely exhausted by their semiotics, or their imbrication with human subjectivity."

CHAPTER THREE

1 Barry's unbounded definition of technozone works well in two ways in this analysis. First, like Elden's (2010) and Mukerji's (2009) work focusing on the active reshaping of earth forms with the explicit aim of establishing dominion over national territory, the technozone makes room for culture and agency, human and nonhuman. Second, so defined, the technozone suggests that a bridge can be crafted between science and technology studies (STS) and the ethnography of infrastructure (e.g., see Hetherington 2019).

2 See Appendix II for a geovisualization of technozones. Flood control is one of two interconnected provincial technozones. The other is owned by the much larger and politically and economically dominant Manitoba Hydro. Following Pritchard's (2011, 20) work on hydroelectric and nuclear industries along the Rhône in France, the technozones produced by MIT and Manitoba Hydro may also be described as "envirotechnical systems" that "may coexist or compete within, or throughout, scales and societies." State power extends, transforming terra into territory, by means of these water-based envirotechnical systems.

3 Vernacular floodways, as I discuss them here, fit into Bourdieu's (1985, 72) definition of habitus:

> The structures constitutive of a particular type of environment (e.g. the material conditions of existence characteristic of a class condition) produce habitus, systems of durable, transposable dispositions, structured structures predisposed to function as structuring structures, that is, as principles of generation and structuring of practices and representations which can be objectively "regulated" and "regular" without in any way being a product of obedience to rules, objectively adapted to their goals without presupposing a conscious aiming at ends or an express mastery of the operations necessary to attain them and, being all this, collectively orchestrated without being the product of the orchestrating action of a conductor.

What I define as blended habitus, then, mixes non-expert folkways of flood-prone inhabitants with science-based technical traditions of forecasting and engineering. Place-based versions of blended habitus, or "infrastructural habitus," I suggest, can be found in any historically flood-prone city (e.g., for expert-inhabitant accounts of Calgary and Toronto summer floods of 2013, see Sandford and Freek 2014). Folkways provide an important dimension of survival that supplement governmental disaster prevention and response programs for floods (Stammler-Gossman 2012; Vaughn 2019; and for tsunamis, Kane 2016). See also notes 1.3 and 1.90.

In this same vein, Whatmore (2013) writes about a "hybrid" project in the town of Pickering, Yorkshire, England, that brings together those with (social) scientific and vernacular flood control knowledge. That project enacts a collaborative space of what I am calling blended habitus. Instead of hybrid, I use the term "blended" to emphasize the mingling of professional and vernacular expertise, a mingling that creates a generative fuzzy zone. (The term hybrid, as in a hybrid engine, often suggests a more mechanical, or at least separable, combination of heterogeneous aspects, domains, or elements.) Whatmore frames the Pickering project with STS, posthumanism, and political theories. For the hydrological approach to the Pickering project, see Odoni and Lane (2010); see also note 8.24. On "socio-hydrology" in Saskatchewan, see Gober and Wheater (2013).

Anand's (2017) *Hydraulic City* illuminates forms of what I am calling blended habitus particular to potable water infrastructure in Mumbai. He finds that "hydraulic citizenship emerges through diverse articulations between the technologies of politics (enabled by laws, plans, politicians, patrons, and social workers) and the politics of technology (enabled by the peculiar and situated forms of plumbing, pipes, and pumps) ... [which] depends on the fickle and changing flows of water" (10).

4 Recurring floods have given urban inhabitants plenty of opportunities to develop sustainable floodways. In his historical review of periodic flood-fighting in the Red River Valley, Bumsted (1997, 9) finds that on the Canadian portion, serious floods have occurred "in 1776, 1826, 1852, 1861, 1916, 1950, 1979, and 1997. Less serious floods came in 1882, 1897, 1904, 1948, 1956, 1966, 1969, and 1974. On the United States portion, where records began only in 1873, major flooding came in 1882–83, 1893, 1897, 1916, 1947–48, 1950, 1952, 1965–66, 1969, 1975, 1978–79, 1989, and 1997." In the Assiniboine Valley, he finds records of major floods "in 1922, 1923, 1927, and 1956." See also Appendix I, Archive and Anthropocene Dialogue.

5 The Water Management Department (https://www.gov.mb.ca/mit/wms/wm/index.html) is within what was called the Ministry of Infrastructure and Transport (MIT) in 2014 during my fieldwork and is now called Ministry of Transportation and Infrastructure (MTI). I continue to use MIT, consistent with publications from that time. For an overview of the Red River (and Assiniboine) basins, see Brooks and Nielsen (2000) and Rannie (2017).

6 For a graphic review of provincial flood-fighting 1826–2013 that contains maps and images of the key structures, see MIT (2013b). For nineteenth-century geoscience description of water bodies' prehistory, see chapter 2.

7 By definition, a floodway diverts water away from a water body, carries it a distance, then discharges it back into the same water body.

8 The 1969 Master Agreement on Apportionment, established on the interprovincial framework established by the Natural Resources Transfer Acts of 1930, was signed by Alberta, Saskatchewan, and Manitoba, three provinces that are vulnerable to both dry and wet cycles, both of which can intensify with climate change. The celebrated multilateral agreement excluded First Nations. https://www.ppwb.ca/about-us/what-we-do/1969-master-agreement-on-apportionment.

9 By definition, a diversion takes water from one body and discharges it into another.

10 For a comprehensive history of rural drains and dikes built by the first settler farmers of colonial times to those governed by modern watershed management regimes, see Stunden Bower (2011) and Haque (2000).

11 I learned this from University of Winnipeg geographer Bill Rannie on the introductory city infrastructure tour he gave to me and fellow geographer Nora Casson on 14 August 2014.

12 Core interviews and site visits with current and retired staff of MIT and Water Survey Canada that form the basis of the basic systems section are listed here in chronological order of first interview with each person (professional affiliations as they were during or before my fieldwork in 2014; field notes are supplemented by photos and drawings): Debi Forlanski (geographer and hydrometric technologist, Water Survey Canada), 3 September and 8 October; Terry Zdan (policy consultant, MIT), 16 September and 7 November; Steve Topping (engineer and forecaster, MIT), with Chris Popper, 2 October, with Darrell Kupchik (director of operations, North Red Waterway Maintenance, Inc., 25 November; Jason Senyk (policy analyst, MIT), 6 October and 10 November; Alf Warkentin (forecaster, retired from MIT), 20 October and 21 November; Rick Gamble (mayor of Town of Dunnattor), 21 October; Doug MacMahon (chief engineer, MIT), 24 October.

Contributing to my holistic understanding of the engineering, ecology, and politics related to this chapter are additional experts involved with water management in municipal or provincial governmental and non-governmental entities and private contractors. Interviews and site visits (with drawings and photos supplementing field notes) include: Kristin Tuchscherer (naturalist, Naturalist Service Office, City of Winnipeg), 2 and 9 September; Jeremy Sewell (geographer and geo-spatial analyst, Municipal Government), 8 September; Dale Hutchison (natural resource manager, Manitoba Hydro), 17 September; Jay Doering (engineer and dean, University of Manitoba), 22 September; Dave Morgan, Roger Rempel, and Seifu Guangul (engineers with climate change and forecasting expertise at Stantec), 3 October; Kristin Koenig and Michael Vieira (engineers, Manitoba Hydro), 9 October; Kendall Thiessen (riverbank management engineer, City of Winnipeg), 10 October and 13 November; Gavin Vanderlinde (mayor of Town of Morris, with Joelle Saltel of NGO EcoWest), 28 October; Myrle Ballard (natural resource expert and Lake St Martin First Nation Community member), 8 November; Major Jeff Bird (emergency manager, Department of National Defence, Canadian Armed Forces, 10 November; Daniel Powell (green project analyst, CDEM, Economic Development Council for Manitoba Bilingual Municipalities, 19 November; John Spacek (urban planner and vice president, CentrePort, 21 November).

Methodological note: It was not easy to gain access to core engineering offices. Once inside, everyone was always gracious and patient, yet I often felt they were looking at me quizzically. In their milieu, the figure of the ethnographer, and the point of ethnography, was no doubt a peculiar one. This was not, however, an obstacle as I proceeded in my aim to figure out how things were supposed to, or tended to, work. For the most part, my research premise – that engineers do the city's geo-cultural work – remained implicit during interviews. To elicit their workaday versions of system design, operational routines, and decision-making protocols, I requested that they make drawings as they replied to my structured questions and engaged in informal conversation. Much of the analytic work in this book proceeded from the technical materials they gave me to take home.

13 Note that in his study of risk assessment and emergency preparedness in the 1997 Canadian flood, Haque (2000, 225) finds "mixed outcomes."
14 Although the infrastructural context is radically changed, people in Manitoba today socially organize activities in accord with the seasons, a "basic ecological adaptation" that Hallowell (1991, 43–59) documents in his ethnography of the Ojibwa based on 1930s fieldwork.
15 See Rashid (2011) for media framing of risk communication and hazard perception in the 1950 and 1997 floods.
16 See Li and Simonovic (2002) with schematic representation of vertical water balance (2647).
17 Alf Warkentin (see note 3.12 for interview details).
18 Aggarwal, Shefali. (n.d.). "Principles of remote sensing," 24, 32–3. Unpublished. www.wamis.org/agm/pubs/agm8/Paper-2.pdf.
19 For Manitoban forecasts, river-level data, and operational status of structures, see https://www.gov.mb.ca/mit/floodinfo/floodoutlook/forecasts_reports.html#about.
20 I include the technologies that my interviewees highlighted. I didn't catch the details of several others.
21 Measured by river gauges set upstream on tributaries.
22 Buzz Currie is the author of the text based on the reports of staff and other journalists, photographers, editors, and a caption writer.
23 On how fragmentation of water management decision-making across river basins with overlapping Indigenous and non-Indigenous jurisdictions, particularly in extreme flood situations, exacerbate problems in Canadian water security, see Canadian Water Security Initiative (2019, 7).
24 For background on Number-Feet James, see Winnipeg City Govern-

ment site, "Flood of the Century: What does James mean?" http://winnipeg.ca/history/flood/james_ave_datum.stm. And Kives (2015), who initiates newcomers. For current river levels at James, see City of Winnipeg Government site: https://winnipeg.ca/waterandwaste/drainageFlooding/riverlevels/current.asp.

25 James Avenue and its pumping station (the origin of today's Number-Feet James) are central to contemporary flood control in the city's history and ongoing revitalization. I include a brief summary from the public archive to suggest the multi-dimensional character of its cultural significance and technical persistence (source list at end).

As it happens, the iconic river measure tied to James Avenue architecture began with a fire, not a flood. The Panama Canal opened in 1903, making Winnipeg's strategic global position in the transcontinental railway system practically obsolete. Although the transoceanic canal would eventually end the big money, in 1904 the Exchange District was still at its peak. So when James Ashdown's hardware store burned down in a raging fire, it was a wake-up call for the commercial-political elites who were invested in the district's new steel-framed skyscrapers. The monied classes imported state-of-the-art pumping machinery from Britain and constructed a building around it. The James Avenue Pumping Station thenceforth drew its water from the Red, and by 1906 it had a role in flood disaster as well as fire management. Someone along the way started the habit of observing and announcing river levels at this stretch of the riverbank.

The immediate environs radically changed around the pumping station through the twentieth century. Immigration waves brought ethnic and linguistic diversity. German, Russian, Polish, Ukrainian, Yiddish, Icelandic, and Romanian entered the English, French, and Indigenous soundscape mix. They had no recourse but to drink from the Red's typhoid-ridden waters while the better-off relied on artesian wells. World War I hit hard, exacerbating workers' suffering. They gathered at nearby Alexander Docks for the General Strike of 1919. Hardship continued through the Depression and World War II. After the flood of 1950 and an estimated 100,000 evacuees, the city built the modern flood control system that would help to change its fortunes. By the turn into the twenty-first century, new modernist and postmodernist architecture joined the Exchange District's beautiful old industrial buildings, which had simply been left waiting for better days through the long economic downturn. And through all of this, until 1986, twenty-four hours a day, the James Avenue Pumping Station provided the city buildings with fire protection. And too, during all these years, someone regularly checked the gauge by the riverbank and recorded Number-Feet James for the populace. Both the pumping station and Alexander Docks, right upstream, are in the centre of Winnipeg's current revitalization.

For history, see Kives and Scott (2013, 38–43) and Virtual Heritage Winnipeg http://www.virtual.heritagewinnipeg.com/vignettes/vignettes_027W.htm#. On revitalization, see Gamble (2019), https://skyrisecities.com/news/2019/05/winnipegs-historic-james-avenue-pumping-station-be-transformed, and Pollock and Bachand (2019), https://www.canadianarchitect.com/on-the-docks-reimagining-winnipegs-waterfront.

26 For a European example of how a focal point is key to adaptation to extreme flooding in the Elbe River Valley, see Albris's (2022, 256) interview of Andreas. Even though his family lives outside the protected urban core, Andreas relies on a focal point in Dresden:

> We now estimate that it takes about ten hours from the first official warning before the river starts to enter our house. We know exactly how much time we have when the water is at a certain level at the measuring station in Elbe.

Recent ethnographies of South American urban flooding document spatialized personal histories. Such practical knowledge, I suggest, could support new communication strategies organized around focal points. Ullberg (2017, 32) discovers "embedded remembrance" in an Argentinean river city. Suburban resident Doña Elena, for example, tells exactly how far up floodwaters reached on an embankment and on the wall of her brick home: "The mark from the water is there, even if you try to paint it over." In a squatter town on the outskirts of Guyana's capital, Vaughn (2022, 5–6) meets a street vendor who "could map the spaces where floodwaters were prone to accumulate." See also Soares (2022).

27 In other words, the establishment of a focal point should be part of the development of "civic infrastructure" (Schoch-Spana et al. 2007) for flood control. Crucially, it is important for blended habitus (or civic infrastructure) to be flexible enough to shift as needed, e.g., with major infrastructural and/or climatic changes. Notably, in 2011 and 2014, when flood extremes shift from Red to Assiniboine, Number-Feet James recedes from public discourse. If this flood pattern persists, a new focal point on the west-to-east axis may arise. See chapter 8 for further discussion.

Note, too, that environmental artists use city-scale focal points as design elements in projects that activate inhabitant water consciousness.

The City as Living Laboratory Milwaukee – Lake Michigan project entitled *Watermarks* by Mary Miss Studios is a great example (see http://marymiss.com/projects/watermarks). Colour-coded lights transform the Jones Island Water Treatment Stack into a "rain forecast indicator to encourage citizens to become part of the green infrastructure of the city, shepherding the use of water before a rain" (37–8). https://static1.squarespace.com/static/53e29b74e4b0d1dab65f7762/t/5d406f417780f10001bc9118/1564503897687/20171117_Short_Booklet.pdf.

28 Once I figured out what the reference to Number-Feet James meant and surmised the probable existence of riverside hardware, I went hunting for the gauge but met with no success, despite help from my sister Claudia and brother-in-law John Michler, who braved the weedy edge. We were in the general area but missed it. Not long after, with a lead from Danny Blair, I met Debi Forlanski, a graduate in geography from the University of Winnipeg, and she took me right to the gauge, hiding in plain sight. (Interview and site visit, 3 September 2014.)

29 Following Manual de Landa (1997), philosopher of the city, Escobar (2008, 11, 35) uses the term "meshworks" for oppositional networks that are "more self-organizing, decentralized, and nonhierarchical" than dominant networks tend to be. I am using meshworks to refer to river–city relationships that may sustain hierarchies and centralization in their midst. These relational meshworks, like blended habitus, fill in open interstices and cross boundaries in riverine technozones. Like Raffles's (2002, 44–5) understanding of Amazonian *varzea*, I understand Manitoban riverine meshworks to be "in the flow of becoming" and, too, in the dash of geo-culture.

30 For analysis of frontstage and backstage aesthetics of secular water infrastructure in Brazil, Argentina, and Singapore, see Kane (2010; 2012, 98–100; 2018a).

31 The transmitter drops into a culvert that goes out to the Red River. As the river goes up and down, it measures water pressure and sends the signal into the control box, which transmits every fifteen minutes via the telephone line to the McPhillips Control Centre. See website references in note 3.25. This is one of thousands of gauges spreading out across the technozones.

32 Calculating stream bed elevation relative to data on river stage is central to flood management. Forlanski explains how she is developing a method for converting the terrain at every gauge station relative to sea level (see figure 3.2). The method avoids having to calculate conversions with every reading, a time saving that could be crucial as rivers peak.

(Second interview at the Red Top Diner, 8 October 2014.) Nevertheless, calculations generally become increasingly dependent on extrapolation during floods, as Odoni and Lane.(2010, 160) explain:

> how does one get access to the gauge when the whole floodplain is underwater, including the gauging station, and when and from where should the water depths and flow velocities be measured? *Clearly more "data" here do not reduce the complexity or uncertainty around the underlying problem – whatever that would happen to be – but extend and complicate matters even more.* [my italics]

See also the technozone map (figure AII.1), which theoretically encompasses the gauge network in all its uncertainties.

33 MacMahon (details in note 3.12). Here is a quote from the Rules of Operation, Appendix A, in MIT 2013a:

> Guideline 1 – Normal Operation
> 1. Maintain natural water levels on the Red River at the entrance to the floodway channel, until the water surface elevation at *James Avenue reaches 7.46 metres (24.5 feet)* or the river level anywhere along the Red River within the City of Winnipeg reaches two feet below the flood protection Level of 8.48 metres (27.83 feet). [my italics]

Note: The term natural refers to the level that would have occurred in the absence of the flood control works, with the level of urban development in place at the time of the design of these works.

34 Dave Morgan interview with Roger Rempel and Seifu Guangul (details, note 3.12).

35 As does Shannon Stunden Bower (2011, 170, and note 1.49) in her study of Manitoba's wet prairielands. Contrast with Anand's (2017, 47–59) ethnographic analysis of the political decision-making chronically affecting the agrarian "outsiders" enrolled in Mumbai's potable water system and the politics of the 2005 monsoon flood disaster inside the hardscaped city (127–30).

CHAPTER FOUR

1 Stuart Elden (2010) defines territory as a political technology that joins ideas of land (property that can be controlled) with terrain (surface over which power can be organized and projected). This definition fits settler colonial practices that pivot on legal control and ownership of property but surely contradicts Indigenous definitions of territory. Indigenous legal scholarship indicates that there might be overlap as well as contradiction in these conceptions. For example, Val Napoleon

(2007, 139) explains that "[l]aw is a process not a thing ... something people actually do." Thinking explicitly about legal order and territory, Napoleon (145) brings in what settlers actually have done by enacting territory as political technology: "The reserve boundaries created by the *Indian Act*, which divided and grouped Indigenous peoples into bands, actually cut across the Indigenous legal orders. This division of Indigenous peoples and lands has undermined the management of the larger legal orders and has undermined the application of Indigenous laws." So plural legal cultures guide the act of joining ideas about "land" and "terrain" in different ways. Exploring the gap between Elden's property-linked territorial definition and Napoleon's plural and processual one could be a useful point of departure for practising "collaborative consent" in the realization of UNDRIP (Simms et al. 2018; see also Napoleon 2009 and Borrows 2019).

2 An earlier version of Part II was published in the edited volume *Territory beyond Terra* (Kane 2018b).

3 Ashley Carse's (2014, 11) ethnographic insight on the Panama Canal is relevant here: "the globe begins to look like a multitude of scale-making projects reaching outward to attach themselves to different infrastructures." Scale-making attempts operate through material and poetic registers that always involve mismatches and maintenance.

4 Site visit, Lyndale Restoration Project, 15 September 2014.

5 Kendall Thiessen, interview and site visit, 10 October 2014.

6 In the Netherlands, managing water levels is not left only to experts. Democratic decision-making about water levels is one of many responsibilities of neighbourhood-level water boards in existence since the thirteenth century (Evers 2019). Farmers generally prefer low levels in winter so they can access polder land with machinery and reduce soil subsidence, but environmentalists prefer to keep them high to maintain vital ecosystems.

7 Steve Topping and Chris Popper, MIT, 2 October 2014.

8 Here I am analyzing the politicality of engineering land that has been made "public" *within* and according to the settler paradigm. There is of course a significant submerged dimension of politicality – river basin colonialism – that is yet to be explored. See note 4.1.

9 Kendall Thiessen, interview and site visit, 13 November 2014.

10 The varied difficulties of managing private property risk in floodplains are becoming ubiquitous in river cities. E.g., for a recent example from the disastrous US midwest floods of summer 2019, see Schwartz (2019).

11 In other words, "Disasters ... create infrastructural tipping points that

make visible infrastructural vulnerabilities" (Kane, Medina, and Michler 2015, 71).

12 In Simonovic and Carson's (2003) technical review of lessons from the 1997 flood, they explain that although the volume of water involved was the same as that in the 1950 flood, the peak was sharper and of shorter duration, and the changes of land use (though incalculable?), "may have also played a role." The river (that became a sea) flooded about 5 per cent of Manitoba's farmland, called up 8,612 soldiers from across Canada, and evacuated 28,000 Manitobans, 6,000 from the city. See also reviews of the 1997 flood by Haque (2000) and Rashid (2011).

13 Photo by Ken Gigioti and caption by Linda Quattrin, assistant city editor, *Winnipeg Free Press*.

14 Rannie (1999) also highlighted overland flooding problems associated with poor zoning and drainage practices on the city infrastructure tour I took with him at the beginning of fieldwork, 14 August 2014.

15 Or, less likely but possible, it might flow into the Assiniboine itself and from there into the city centre.

16 A geostationary satellite orbits around the earth at the same rate that the earth turns. An observer at any place where the satellite is visible will always see it in exactly the same spot in the sky, unlike stars and planets that move continuously.

17 John Law's (1994) argument in *Organizing Modernity* is consistent with their efforts. Law argues that carrying out "plural processes of sociotechnical ordering" that are composed of conflicting "materially heterogeneous" practices sets the stage for reproducing social order. Applied to flood control, this characterization suggests that the better designed the flood control system – in Mukerji's terms, the greater the logistical power it offers – the more it will contribute to the strategic power of the city dwellers who benefit. Law's is not an overly deterministic theory, i.e., there's room in it for causality to be multi-dimensional and even unspecified. It fits well with Barry's (2006) technozone concept (see Appendix II). Compare Anand's (2017) ethnography of Mumbai's water engineering practices.

18 Engineers have a healthy respect for the dynamic complexity of Winnipeg's flood control system in the intersection of two river basins. In my interview with Doug MacMahon, before addressing the series of questions I presented beforehand at his request, the first fact that he wanted me to understand was that there are many different kinds of flooding. His point counters a general cultural predisposition (to which I had fallen prey) to focus attention on how MIT makes operational deci-

sions regarding the monumental structures – the Floodway, the Primary Dike System, the Portage Diversion – and how they affect the Red and Assiniboine so as to alter levels at Number-Feet James. But the Red River Valley is a fantastically complicated terrain with many sub-basins, towns, farms, industries, entwined infrastructures (flood control, transport, hydroelectricity, overland drainage, sewage), each with its own governance regimes, inter-relationships, and, as highlighted in this chapter, cyber-earthwork systems, seasonal routines, and unintended consequences. Thanks also to Terry Zdan for bringing these matters to my attention. (Interviews 16 September and 7 November 2014.)

Geo-cultural understanding requires a good grasp of how infrastructures mediate human relationships with rivers. But an appreciation of, for example, interlaced flooding regimes illuminates the general throwntogetherness that is essential for empirical honesty. Experts produce sound analyses within limited frameworks by simplifying complex phenomena. Indeed, modelling complexity demands simplification. However, without honest consideration of range of throwntogetherness effects, engineers cannot build resilience into systems.

CHAPTER FIVE

1 The focus on lake levels in blended habitus exemplifies how a technicality also organizes sociality. Lake levels simultaneously act as water body features, data, law and policy, and infrastructure, as well as cognitive, emotive, and discursive focal points. As abstractions with various physical manifestations, lake levels are the building blocks of cultural repertoires in negotiations about how unintended or unavoidable harms might be prevented or made right. For a Brazilian case, see "Sense and Science at the Lake of Dark Waters" in Kane 2012, 23–40. Contrast Simpson's (2008) analysis of balance in the Anishinaabe concept *minobimaadiziwin* (the good life); see Simpson 2008, 32, chapter 1, note 1.59, and Ballesteros's (2019) analysis of how ideals of balance and harmony guide regulators who calculate the price of potable water in Costa Rica.

2 Referring to water as simply an elemental collection of molecules, H_2O, reduces it to an abstraction, thereby erasing most of its social powers (Linton 2010). However, water's embodied animacy is elemental, changing with phase and state. For example, once ice or liquid water evaporates as vapour and disperses into air, a human's relationship with water-as-lake changes and becomes more difficult to discern (see Adey 2014, 15).

3 To clarify why forms of embodiment are essential in the ethnography of flood control, consider the contrast between two interpretive guidelines used by Manitobans as flood approaches. For rivers, Number-Feet James (river level at its intersection with James Avenue) indexes the combined force of the Red and Assiniboine Rivers (see chapter 3). Any given reading must be interpreted relationally: inhabitants adjust their understanding so that they can figure out what might happen in their own diverse positions within intersecting floodplains. In contrast, the unifying and orienting measure of lake level does not require a relational form of diverse interpretation because water flowing into lakes spreads out. All things being equal, lake waters flow in such a way as to achieve the same level across their water body surfaces. Of course, all things are never quite equal. Forces such as wind, the weight of ice, human-engineered control structures, and untimely events confound lakes' equalizing action and its level reputation for inspiring equanimity. (Aquatic ecologists distinguish between "lentic," still waters of lakes, and "lotic," the running waters of rivers. It seems an over-simplification, especially in floods and storms, when winds whip up waves in lakes.)

4 Lake levels are also important measure in Manitoba Hydro's regulation of Lake Winnipeg but in a different way than MIT's regulation of Lake Manitoba. (See technozone map in Appendix II.) The contested lake-level range within which Manitoba Hydro may regulate Lake Winnipeg levels is governed by LWR (Lake Winnipeg Regulation) of the Water Powers Act. In informal conversations with a few people on the shore in Victoria Beach and Grand Beach in Lake Winnipeg early in my fieldwork (20 August 2014), I learned that they felt strongly that Manitoba Hydro sacrificed their security in floods in order to maximize electricity production. So when I met with Dale Hutchison, the Waterway Community Engagement section head in the Winnipeg office of Manitoba Hydro not long afterwards, I asked him about this (interview, 17 September 2014). He took great pains to assure me that although people affected by flooding on Lake Winnipeg believe Manitoba Hydro can and does "control" lake level, it cannot; in fact, it can only "regulate" lake level. Hutchison also explained that flooding is not especially good for hydroelectricity production. Efficiency requires a steady, even flow through the step dams linking Lake Winnipeg to Hudson Bay. Manitoba Hydro's system does receive floodwaters from Lake Manitoba and Lake St Martin, which sit higher in the landscape and flow down through a network of "flat and meandering rivers". See

Section 5 of the 2014 report: https://www.hydro.mb.ca/docs/regulatory
_affairs/pdf/lake_wpg_regulation/lwr_complete_report.pdf.
5 For structure of feeling, see note 1.90.
6 Scott Forbes (interview and site visit, Twin Lakes, Lake Manitoba, 12 August 2014; follow-up interviews, 24 September and 23 October 2014). Forbes shared his invaluable, critical perspective on flood control; his insights on wind and water inform the analysis of WEE below. Thanks to geographer Danny Blair for making the connection and for sending me the links to Forbes's many post-2011 *Winnipeg Free Press* letters.
7 Albeit more carefully, chief engineer Doug MacMahon confirms this fact (interview, 24 October 2014). See chapter 3 for more in-depth discussion of system routines and rules.
8 Forbes also argues that ultimately, engineers have no solution for the ecological cost of their continual overuse of the diversion, which eventually will be catastrophic. In support of this contention, he offers an example that hints at wind's complex role: as dangerous as strong northwest winds across Lake Manitoba can be, winds within normal range are essential to lacustrine ecology. Only wind keeps the lake waters oxygenated. Recently, every time the wind dies down, an algal bloom causes a massive fish kill. The ecology is off-balance for multiple reasons, including river engineering.
9 The photos and text on the Wikipedia page for Twin Lakes Beach capture the before and after of the 2011 flood disaster. The references include government reports and newspaper articles, as well as Forbes's letters in the *Winnipeg Free Press*. https://en.wikipedia.org/wiki/Twin_Lakes_Beach,_Manitoba.
10 See chapter 4 (110–11) for an explanation of overland flows.
11 The field station of the Delta Marsh project sustained so much damage in the flood of 2011 that the University of Manitoba shut it down.
12 The state performs public accountability by disseminating ribbons across the landscape. This performance at the elemental interface between water and land is another example showing how mundane technical decisions can become a focal point organizing debate in the media, the streets, and the courts. They transcend their own technicality as they penetrate sociality (Kane 2018b).
13 Inhabitants of both Twin Lakes Beach and Lake St Martin were subject to evacuation orders, the latter in more dire circumstances. In Lake St Martin, people were given twenty-four hours to evacuate a place that had been inhabited for more than 140 years. Half refused to evacuate, the other half was displaced and scattered across the province. There are

an additional three First Nation communities on Lake St Martin: Dauphin River, Little Saskatchewan, Pinaymootang (see maps in Thompson 2015). Ballard and the Interlake Reserves Tribal Council (IRTC) gathered elders from the four communities in 2015 as part of the Manitoba Flood Healing Project, Minoayawin. http://manitobaflood healingvoices.com/index.php/the-project.

14 Myrle Ballard, interview, 8 November 2014.
15 Ballard (2012) got her PhD in natural resources from the University of Manitoba. She wrote a thesis on language, gender, and the flooding of sustainable livelihoods in Lake St Martin First Nation. She made the 2012 film, *Flooding Hope*, with Shirley Thomson, co-writer and producer, and Ryan Klatt, director of photography and editor. Ballard also wrote, directed, and co-produced a video about the 2011 evacuation of another Lake St Martin community entitled *Wounded Spirit: Forced Evacuation of Little Saskatchewan First Nation Elders* (Ballard, Martin, and Thompson 2016).
16 Strathern 1999, 221, cited in Jiménez 2017, 77.
17 See proposed international legal definition of ecocide at https://www.stopecocide.earth.
18 In the title story of *The Brothel Boy*, Norval Morris (1992) illuminates the concepts of blame and guilt in public understanding of legal judgment.
19 Note that engineering's idea of nature skips over Indigenous habitation and engineering practices from time immemorial.
20 See also chapter 4 for how wind unknowns shaped suspense at the Z-dike in the flood of 1997.
21 On balance, see, e.g., Simpson (2008, 32, cited in Craft 2014, 6–7). For an example of how critical settler cartographies can draw on relational Indigenous concepts, see also Fujikane's (2021, 16–17) discussion of the Kanaka Maoli (Native Hawaiian) concept of *pono* (just, balanced, and generationally secure).
22 C.C.S.M. c. W70. The Water Resources Administration Act. Manitoba 2018.
23 The 1977 hydrology map created by Canada, Fisheries and Environment, Prairie Provinces, shows active hydrometric stations clustered across the southern portion of the province in the Assiniboine and Nelson River basins. There are practically no stations, and thus no data collected, in or for the benefit of the far north (Weir 1983, 17).
24 Naomi Oreskes (2007) finds that the task for modelling, such as hydrological modelling, is to generate predictions that can inform policy decisions so that, e.g., provincial governments have some basis for say-

ing yes or no to questions posed. No one actually expects the models to be able to capture future geophysical reality as such. She argues that when geologists engage in predicting the future, their "role is primarily social, not epistemic." Applying Stengers's (2010) philosophical provocations to a river-city flood management intervention, Whatmore (2013, 9, 45), in contrast, argues that the tactic of forcing a "slow down" of expert reasoning opens hybrid spaces for reasoning differently. I suggest that both Oreskes's and Whatmore's arguments are applicable to the effort to evaluate expert-inhabitant action in past floods and in ongoing disaster management communication in the meso (see chapter 7) and as "mesopolitics" (Stengers 2008).

25 For Assiniboine hydrogeology, see also chapters 2, 3, and 8.
26 MIT 2013b, 52–5. Significantly, the Fairford River Control Structure was designed for 25,000 cubic feet per second (cfs), but in the flood of 2011, the engineers sent 34,000 cfs its way. Both Topping and Warkentin, engineer and forecaster, emphasize this point in their interviews with me, adding the 2011 cfs numbers to their impromptu drawings of the flood control system. (Interviews of Steve Topping with Chris Popper, 2 October 2014; Alf Warkentin, 20 October 2014).
27 Definition in MIT (2013b, 77, note 2) "dam^3 stands for cubic decametre, which is equal to 1,000 cubic metres; 1 dam^3 is equal to 0.81 acre-feet; 1 dam^3 of water would cover a square kilometre of land to a depth of 1 mm."
28 This is Scott Forbes's argument on the chronic use of the Portage Diversion. It also applies to Lake Manitoba's north end.
29 For examples of how the Indian Act affects Indigenous scholarship and peoples more generally, see Napoleon (2007, 139, 145), Ballard (2012, 136–9), Coulthard (2014, 83–91), Hogue (2015, 15), Razack (2016, ii), and Liboiron (2021, 32, note 128).
30 Internal frontiers, I suggest, could be considered hallmarks of anthromes, or anthropogenic biomes, if environmental scientists seriously consider how infrastructural incursions have reshaped the areas they call "wildlands" (Ellis and Ramankutty 2008). Describing the wider context of human-environment research, Brondízio and Chowdhury (2013, 393) write that, among other things, infrastructural expansion and urbanization underlie an "evolving matrix of social-territorial complexity worldwide." Discovery and analysis of internal frontiers that implicitly distort the spatial justice dimensions of infrastructural expansion and urbanization could be part of human-environment research.
31 See Thompson (2015) for a legal history of Manitoban water management focusing on the flood of 2011 as yet another case of Indigenous

dispossession. The import of her counter-narrative to the MIT report (2013b) is consistent with this chapter's critical reading and goes into detail about the construction and operation of the Portage Diversion, the Fairford Water Control Structure, and the emergency channel.

32 Cameron, de Leeuw, and Desbiens (2014, 24) articulate the danger of a reawakened *terra nullius* lurking in contemporary academic approaches to traditional Indigenous knowledge: "Without explicitly interrogating industry and government efforts to define and clearly locate Indigeneity, the move toward recognition will instead work to 're-establish a *terra nullius* open again to development but mildly constrained by discrete, localized, patches of Indigeneity.'"

33 As a state-driven territorializing tactic, internal frontiers may have diverse geo-political dynamics. For example, in his history of the Itaipu Dam flood, Blanc (2019, 180–1) finds that the dictatorship sent dispossessed farmers who held no land titles into practically unlivable "backlands" in order to strengthen Brazil's claims over a distant "internal frontier."

34 And thus, Coulthard (2014, 175–6, cf. Smith 1996 among others) argues, dispossession of Indigenous lands occurs both within cities, where more than half of Canada's Indigenous people live, as well as in the hinterlands. Flood control, like gentrification, is not simply a necessity or improvement but, rather, state projects central to the accumulation of capital. On this theme applied to Winnipeg, see Tomiak et al. (2019). For analysis of the inside/outside logic of racialized spaces and marginalization through flooding in Manitoba and globally, see Gill (2000, 141–2, 147, 151).

35 Internal frontiers foster an atmosphere of legal withdrawal wherever jurisdictional ambiguity provides an alibi (as when First Nation sovereignty appears to create complications that block provincial and/or federal reconstruction efforts). In the flood of 2011, resolving the question of which governmental entity should be responsible for emergency flood preparedness for and recovery of artificially flooded First Nation communities was stymied by fuzzy boundaries between provincial and federal governments and the unrepealed yet reviled Indian Act. See Shrubsole (2001) for an analysis of Eurocentric cultural bias and jurisdictional obstacles in the 1997 flood. See Ribeiro (1994) and Kane (2012) for analyses of intentional formation of jurisdictional ambiguities in internationalized aquatic landscapes by corporations and regional governments in South America.

36 The spatio-temporality of flood control systems contrasts with that of

Manitoba Hydro's step dams linking Lake Winnipeg to Hudson Bay (see figure AII.1 in Appendix II). Unlike MIT, Manitoba Hydro is a larger, profitable system, and its governance is semi-autonomous. Yet despite differences in scale, function, and autonomy, both technozones rely on flooding First Nation territories and displacing communities for the benefit of the city. See Wa Na Ski Tan, "An Alliance of Hydro-Impacted Communities," http://hydroimpacted.ca/wa-ni-ska-tan-2. And too, both flood control and hydroelectric production systems embed internal frontiers inside the prominent space of the legalized province.

CHAPTER SIX

The idea for the term "sunlight machines" comes from "our best machines are made of sunshine," Bennett and Chaloupka (1993, xi), citing Haraway (1985).

1. In this *New Materialism* interview, Barad (2012, 53) goes on to say, "This is a feminist project whether or not there are any women or people or any other macroscopic beings in sight." In this spirit, I write about diatoms moving through categories and disciplines. See also *Material Feminisms* (Alaimo and Hekman 2008).
2. Relevant here is Braun's (2000) study of the development of nineteenth-century geological ways of seeing, particularly as applied to mining, and how geologists influenced political rationality rather than being a mere instrument of it.
3. See also Tsouvalis, Waterton, and Winfield (2012), who build on Barad and Latour to carry out a community-based study discerning how scientific data about microscopic algae come to matter, materially and politically, in the negotiation of uncertainty.
4. Technically, Lakes Manitoba and Winnipeg are fluvial lakes because they were created by rivers meandering through floodplains (see chapter 2). On the legal and metaphorically significance of hydrological balance and lake level, see chapter 4.
5. Following dominant scientific tradition, textbook authors construct geophysical facts by omitting human actors and their earthworks. They write as if Method allows physical geographers to separate themselves and their social lives from the (quite dead, or at least insensible) objects they study. The style banishes its creators, their relationships, and influences in the name of objectivity, reproducing the material imaginary that the public has come to expect from experts. It is why feminist scholars (e.g., Haraway 1988) have been compelled to unpack the

untold social influences in the production of facts, inventing and reinventing Feminist Science and Technology Studies. And too, why social and cultural anthropologists and their allies (e.g., Clifford and Marcus 1986), influenced by feminist critiques of objectivity in science and social science, turned toward more reflexive forms of ethnography. However, I must say, it's a lot more difficult figuring out how to be reflexive when collaborating with nonhumans.

6 Email exchange with J.P.M Syvitsky, 14 April 2018.
7 I have drawn from a variety of sources to write about diatoms, including most prominently Round, Crawford, and Mann (1990), Smol and Stoermer (2010). I find that the introductory and method portions of scientific publications are mostly likely to provide data of cultural relevance. The Stoermer, of Smol and Stoermer (2010), is the very same scientist who published the two-page Crutzen and Stoermer (2000) piece entitled "The Anthropocene" in the International Geosphere-Biosphere Programme (IGBP) *Global Change Newsletter* that set geoscience on fire. He and Smol do not use the term in their diatom tome, but they do capture the Anthropocene spirit when they write "[it] becomes increasingly evident that human actions are exercising greater control over the conditions and processes that allow for our existence" and "the study of diatoms as tools that can be used to infer the direction, magnitude and limits of change becomes increasingly imperative (4)."

For work on diatoms designed for the public, see "Diatoms of North America," https://diatoms.org, and the California Academy of Sciences (n.d.), "Introduction to Diatoms," http://researcharchive.calacademy.org/research/diatoms/overview/introduction.html.

8 For global sustainability implications of diatoms and their connection to the animal-like urea cycle, see European Commission 2011.
9 Julius and Theriot (2010, 18) do not refer to "specialized teeth of mammals" randomly. Indeed, it relates obliquely to a 1732 insight that led Linnaeus to invent the system of nomenclature in use today (Blunt 1971, 57). It is, I believe, worth considering the great breadth of morphological exclusion upon which this founding insight relies (Kane 2019).
10 This makes me wonder if diatom studies might figure in Hustak and Myers's (2012, 79, 97, 106) project, as cited by Donna Haraway (2016, 68) in *Staying with the Trouble*:
   a theory of ecological relationality that takes seriously organisms' practices, their inventions, and experiments crafting interspecies lives and worlds. This is an ecology inspired by a feminist ethic of

"response-ability" ... an affective ecology in which creativity and curiosity characterize the experimental forms of life of all kinds of practitioners, not only humans.

11 Their metazoan-like capability adapts diatoms to episodic nitrogen availability and is essential for marine productivity.

12 Calling on earlier transdisciplinary border crossings, Bennett draws on the "biogeochemistry" of Russian scientist Vernadksy (1863–1945) who, refusing the dichotomy of life and death, preferred to speak of "living matter," of life as a happening or process, rather than a thing. See Vernadsky ([1938] 2001). Zalasiewicz et al. (2017) also draw on Vernadsky to develop the concept of the technosphere (see also Appendix II, note 1).

13 Shallow lakes are more vulnerable to a given pollutant load than large lakes (Bennion et al. 2010, 152). Responding to requirements of environmental legislation such as the EU Water Framework Directive, scientists have developed diatom assemblage as biomonitoring tools, as indicators of the amount of biological activity a lake can sustain (trophic status) (194).

14 In *Friction*, Anna Tsing (2005, 262) writes: "What if we paid more attention to *incompatible* data sets? ... Instead of erasing incompatibility, we need to find out where it makes a difference."

15 Eva Pip interview, 14 November 2014.

16 Theirs is an international and interdisciplinary relationship. Jan Risberg is in the Department of Physical Geography and Quaternary Geology, Stockholm University; Hedi Kling is from the Department of Fisheries and Oceans in Winnipeg's Freshwater Institute; and Hannelore Håkansson is from the Department of Quaternary Geology, Lund University. See note 5.11 about the Delta Marsh Field Station.

17 Light microscopy is the only way to see living diatoms. But using electrons instead of light beams to create images affords better resolution at high magnification, greater depth of field, and allows microanalysis of material composition. The drawback of SEM is that the objects must be rendered conductive: Håkansson and Kling coated their samples with gold. http://www.ammrf.org.au/myscope/sem/background. SEM (= scanning EM) images are three-dimensional, whereas those of TEM (= transmission EM) are two-dimensional.

18 This is another example of the way that invention of different visual technologies pushes the depth and range of data collection in ways that may significantly affect understanding of deep time events (see chapter 7).

19 These included light microscopy, mineral magnetic analyses, paleo-salinity reconstructions, and radiocarbon dating.
20 In their summary, Bennion et al. (2010, 167) include the influence of the EU Water Framework Directive. Adopted in 2000, this legislation emphasizes the ecological status of water bodies. The EUWFD has "Renewed interest in the role of diatoms in understanding shifts in ecological structure – especially as a component of multiproxy studies which seek to determine whole-ecosystem responses to environmental change."
21 Murchie's book comes to me by way of my late father, Bernard David Kane (1922–2021), who read it as a young meteorologist and who continues to inspire my interest in wind.

CHAPTER SEVEN

1 Assiniboine flooding is certainly not unprecedented. See note 3.4 for dates of historical flood extremes.
2 A fact emphasized in interviews with private and public sector engineers Roger Rempel (3 October 2014) and Doug MacMahon (24 October 2014).
3 The performance of political engineering is another example of what John Law (1994) calls the "plural processes of socio-technical ordering" composed of conflicting "materially heterogeneous" practices that set the stage for reproducing social order.
4 Bakhtin (1981, 84) borrows the idea of chronotope for literary criticism from Einstein:
> We will give the name *chronotope* (literally, "time space") to the intrinsic connectedness of temporal and spatial relationships that are artistically expressed in literature. This term [space-time] is employed in mathematics, and was introduced as part of Einstein's Theory of Relativity. The special meaning it has in relativity theory is not important for our purposes; we are borrowing it for literary criticism almost as a metaphor (almost, but not entirely). What counts for us is the fact that it expresses the inseparability of space and time (time as the fourth dimension of space).

5 Thanks to Danny Blair for bringing Manitoba's "new normal" and "non-stationarity" to my attention in 2014.
6 Data-based models of attribution studies verify the probability that intensifying floods are due to anthropogenic influence rather than natural variation. For models at North American continent scales, see

Kirchmeier-Young and Zhang (2020). See Sauchyn, Davidson, and Johnston (2020) for a literature review of climate change impacts and adaptations in the Canadian prairies.

7 For video of and reporting on the press conference, see Owen and Kusch (2014). In discussion below, I put the time of the statements in brackets, e.g., [1.00] indicates that the statement can be found at approximately one minute into the videotape. For my full, rough transcription, see Appendix III.

8 The most vulnerable subjects of this press conference are those in the projected inundation zone of the Hoop and Holler dike breach and the communities and farms on either side of the insecure dikes between the cities of Portage la Prairie and Winnipeg, including the city of Brandon.

9 The rhetorical mode of emergencies provides a counterpoint to the moderating impact of daily weather-reporting, which presents the broader context of extreme flood events. In *North of Empire*, Jody Berland (2009, 219–20) writes:

> Central heating, air conditioners, climate-controlled cars, underground malls and pathways, and weather forecasts participate in a process of technical mediation which teaches consumers to structure and manage routines of everyday life with ever greater control over spatial and temporal conditions.

The press conference partakes of this "cult of efficiency" yet functionally intervenes in river–city relations in a way that requires different organizational tactics on the part of government.

10 Following Philippopoulos-Mihalopoulos (2015, 7–9) on spatial justice, one could analyze the Hoop and Holler situation as a withdrawal of provincial government protection from Hoop and Holler, a withdrawal that makes room for the river's impulse to burst out of its channel and spread over the prairie. Drawing on the concept of conatus, or "the will of each body to carry on being and becoming," Philippopoulos-Mihalopoulos argues that "while always part of assemblages with other bodies, every body withdraws" (8). His is a welcome complication of both Latour's Actor Network Theory and much of Science and Technology Studies, "which do not reserve a space for non-connection" (9). Indeed, withdrawal of protection, as much as extending protection, should be counted among one of Law's "plural processes of socio-technical ordering" (see note 7.3) that shore up provincial authority. See also my discussion of internal frontiers in chapter 5. Moreover, withdrawing from state-imposed spatial justice can be a form of resistance (as in the Black Lives Matter movement's goal to defund the police).

11 While their terrain and material practices are quite distinct, a similar

argument can be made for Manitoba Hydro. (See Appendix II on technozones.)

12 Flipping the order of cause and effect makes visible more sinister possibilities, disturbing the axis of trust between officials and citizen-majorities. In his comprehensive discussion on mapping space, Langdon Winner (1977, cited by Berland [2009, 257–8]) brings attention to the process he calls "reverse adaptation." What if, rather than building an assemblage of technological artifacts – e.g., those linking river gauge data to press conference performance for the purpose of managing river floods – the military first builds the artifact assemblages and then looks around for possible applications, one of which happens to be flood control? How might this kind of reversal impact governmental priorities?

13 See note 7.7.

14 Mobilization of army personnel and technologies for flood defence and forecasting blurs the civilian-military divide. Such discourse is carefully policed to assure the public that the military is under civilian control at all levels. (Interview, Major Jeff Bird, 10 November 2014.)

15 The press conference has meaning only in the context of the timescape of the unfolding flood. In her research on disasters, Barbara Adam (1998, 11) invents the timescape concept as "a way of seeing and a conceptual approach that shifts emphasis in environmental praxis from explicit space and implicit time to the complex temporalities of contextual being, becoming and dwelling." Combining Adam and Bakhtin (1981, 250), then, in the chronotope the actors performing the short press conference pull the larger timescape into narrative, "assimilating and appropriating," condensing and mobilizing its suspense.

More generally, Adam's concept of timescape frees ethnographic analysis from the stasis that underlies the language-place connection practised by Basso (1996). The concept of timescape (like Barry's concept of technozone) makes room for indeterminacy and works well with transdisciplinary approaches to elements, such as water and earth, as interactional, or intra-interactional (Barad), and transformative (see, e.g., Peters, Steinberg, and Stratford 2018). Thanks to political geographer Phil Steinberg for his critical insight into the implications of Basso's foundational work in anthropology.

16 At the same time, the Manitoba Flood Outlook reports: "There is overbank flooding in the Assiniboine valley in all reaches between the Shellmouth Dam and Brandon." http://www.gov.mb.ca/mit/floodinfo/floodoutlook/forecast_centre/daily_reports/2014/en/flood_report_july_4_2014.pdf.

17 No matter how much data the government has access to, thunderstorms can always introduce unpredictability. For an island example, see Kane (2018a) about why flash flooding stands alone as a kind of event that the Singaporean government cannot control in its hydrohub.
18 The camera does not swivel to capture images to accompany questioners' voices.
19 See provincial media bulletins: http://news.gov.mb.ca/news/index.html?archive=&item=31745 and http://news.gov.mb.ca/news/index.html?archive=&item=31771.
20 The actual crest on 9 July reached 50,000 (not the predicted 52,000), raising the question of whether MIT was being overly cautious at the expense of the lake dwellers.
21 See Kane, Medina, and Michler (2015) about how lack of scientific understanding about anomalous edge waves led to the failed tsunami alert of 2010 and its major political consequences on the Chilean national stage.
22 Structural engineer Roger Rempel questions the continued reliance on traditional flood pattern parameters in building codes and on governmental bidding processes that favour the lowest-cost bids rather than those that strive to account for the wider range of flood extremes expected with climate change. (Interview of Roger Rempel with Dave Morgan and Seifu Guangul, 2 October 2014).
23 Bakhtin (1981, 243) writes: "[In] ... the chronotope of encounter ... the temporal element predominates, and it is marked by a higher degree of intensity in emotions and values ... spatial and temporal series defining human fates and lives combine with one another in distinctive ways, even as they become more complex and more concrete by the collapse of *social distances* ... a point of new departures and a place for events to find their denouement."
24 For Williams's (1977, 132) definition of "structure of feeling," see note 1.90.
25 Bakhtin (1981, 243).
   [H]ighly charged with emotion and value a chronotope of threshold ... can be combined with motif of encounter, but its most fundamental instance is as the chronotope of *crisis* and *break* in a life ... In this chronotope, time is essentially instantaneous, as if it has no duration and falls out of the normal course of biographical time ... becoming part of mystery ... and carnival.
26 In the beautifully illustrated *Flood: Nature and Culture*, John Withington surveys the prominent role of the flood in literary and visual arts,

myths, and religion, with an emphasis on Western culture. His chapter "Description: Floods in Literature" (59–81) surveys literature beginning with a children's book, *Hans Brinker, or the Silver Skates*, by Mary Mapes Dodge (1865) that includes the story of how the eight-year old Dutch boy saves his village by keeping his finger in the dike all night long. Faulkner's *The Old Man* is prominent and, unsurprisingly I suppose, includes dramatic pieces from the selection I present here. Withington's survey reaches into forerunners of today's climate change fiction ("cli fi") with novels such as J.G. Ballard's *The Drowned World* (1962) and Stephen Baxter's (2008) *Flood*. In addition to Withington's book, ancient myths including the biblical Noah are discussed by Dundes (1988) and Da Cunha (2019, 205–7). For an example of how Indigenous people, the Emberá of the Darién rainforest in Panama, assimilate the biblical flood myth into their own place-based flood histories and oral story-telling tradition, see Kane (1994, 33–8).

27 Elyachar analyzes how gossip creates channels for economic infrastructure. See also Hoffman's (2020, 279–87) discussion of the multi-dimensional importance, often ignored, of local expressive cultures in disaster prevention and response and the volume as a whole for an in-depth collection of disaster anthropology.

28 Thanks to Beverley Peters for telling me about Ian Ross's play. (Interview and site visit to Omands Creek, 9 November 2014).

29 My analysis is based on the published script (Ross 2001). *The Gap* premiered at the Prairie Theatre Exchange in 2001 (Canadian Theatre Encyclopedia, https://www.canadiantheatre.com/dict.pl?term=Ross%2C%20Ian). The title of the Cree version is *Towaw*, translated by Joyce Noonen. (Manitoban Association of Playwrights, http://mbplays.ca/plays/the-gap).

30 See also Tomiak et al. 2019.

31 Fransje Hooimeijer (2014, 6), who writes about a "Fine Dutch Tradition" in polder cities, articulates the sociopolitical benefits of living in a flood-prone landscape:
> Social behavior determines history within the given conditions of the region. In the Dutch case the wet and soft soil conditions created a collective enemy resulting in a strong feeling of citizenship and a powerful tradition in water management expressed in the very early installation of a public body in the form of the Water boards [the nation's oldest democratic institution].

Hooimeijer (2014, 11–12) identifies the historical phases of urban development by studying the interfaces of architecture, infrastructure, and water management. She identifies and defines four phases of Dutch polder cities layered into the built environment of the present: I. Natural water state (until around 1000); II. Defensive water state (1000–1500); III. Offensive water state (1500–1800); IV. Manipulative water state (1800–present). The framework could usefully be applied cross-culturally, adjusting the dates to fit the shifts in what I call blended habitus in Manitoba and elsewhere. In her systematic overview, Hooimeijer (244) calls phase I and II "amphibious culture," phase III and IV "land culture," and projects a sustainable "resilient culture" for the future. (See chapter 1 application of her terms to Manitoba and note 4.6 on water boards.)

32 Bill Rannie (Interview and infrastructure tour, including the Floodway, 14 August 2014).

33 Bakhtin (1981, 250) summarizes:

Thus the chronotope, functioning as the primary means for materializing time in space, emerges as a center for concretizing representation, as a force giving body to the entire novel [or play]. All the novel's abstract elements – philosophical and social generalizations, ideas, analyses of cause and effect – gravitate toward the chronotope and through it take on flesh and blood, permitting the imaging power of art to do its work. Such is the representational significance of the chronotope.

34 This reversal of analytic method is inspired by Marilyn Strathern's (1992, 3, cited in Helmreich 2003) thinking about continuity and change: "Instead of thinking what they measure, we might think about how each depends on the other to demonstrate its effect." Compare Whitington's (2018, 224) ethnography of a hydropower company in Laos in which he finds: "Uncertainty is not the opposite or lack of certainty; rather, it is inherent in the practice of knowledge."

CHAPTER EIGHT

1 Rephrasing in Barad's (2010, 268) terms, what happens when engineers face the "ghostly causality" of rivers?

2 In conversation, Bill Rannie said that before 2011, the Assiniboine was a "sleepy little prairie river, anonymous" but that it had the distinction of being the only prairie river that is only fed by the prairie. Bill Rannie, Geomorphology Field Trip, 21 September 2014. For details of Assiniboine River history, see chapter 2; for floods of 2011 and 2014, see chapters 4 and 6, respectively.

3 The *Oxford English Dictionary* entry for avulsion skips right to the

geopolitical; geomorphology is implicit. The first definition pertains to sovereignty: "The action of pulling off, plucking out, or tearing away; forcible separation." In the first example, from 1622, seawater is the space that enables territories to be forcibly separated (Sicily from Italy, Cyprus from Syria, the United States from Britain).

Geomorphologically, key features of avulsion as a mode of aquatic agency are speed and displacement. These features carry over into contemporary Canadian law wherein a sudden and "perceptible action of water" serves as a precondition for disputes between property-owners or states that experience a loss in terra firma, a quantity of soil, or the certainty of a boundary. In such cases, the law sets forth the principle that "the resulting change of channel works no change of boundary" (Yogis 1998). Thus, in the context of Canadian legal disputes, water's nonhuman agency is neutralized.

4 Natural channel migration occurs everywhere. Resisting migration through riverbank maintenance is a regular responsibility of city government. It entails tasks that are more feasible on public than on privately owned riverbanks. See contrast of Red River neighbourhoods Norwood and Riverview in chapter 4.

5 Bill Rannie, Geomorphology Field Trip, 21 September 2014.

6 See chapters 1 and 5.

7 LiDAR's three-dimensional images with high spatial resolution fill in the sculpted elevations of the lower Assiniboine's "unusual geologic history" more precisely than those based on aerial photography alone. An unnatural color code indexes areas of different soil composition of the surface terrain through which the Assiniboine flows. Other technologies are relevant to discerning Assiniboine migration history. E.g., Bater (2002) collects samples from alluvium for radiometric dating of *Bison* (Linnaeus) bones and for dendrochronological dating of oak logs (subfossil *Quercus macrocarpa* [Michx.]). Thanks to Jason Senyk, policy analyst, for giving me a copy of the 2011 *Flood Technical Review* (MIT 2013b) and alerting me to the significance of the LiDAR images. Interview at MIT, 10 November 2014.

8 A river flowing in its familiar channel can also encounter itself in/as a different-state-of being. This is another dimension of body multiple. For example, in spring, when the Red flows north from the Dakotas as liquid, it may meet vertically growing frazil ice. The frazil ice forms chokepoints at river bends and bridges and can trigger outbursts. See chapter 3.

9 Smithson continues: "Theories like things are also abandoned. That the-

ories are eternal is doubtful. Vanished theories compose the strata of many forgotten books." The artist is writing about the theory that connects a photograph of his earth art in a gallery to the actual earth work in the New Jersey Pine Barrens. In our geo-cultural moment, it seems, the theory that empirically connects LiDAR representations of an always-possibly-avulsing Assiniboine seems to carry a similar sense of tentativeness. The relationship between gallery artifact and earth-art-thing will have to be revised, as will the relationship between the geoscience corpus and engineering practice. And in this ethnographic terrain, Smithson, the artist, meets Massey, the geographer (see Facts, Feelings, and Throwntogetherness, chapter 1, part IIC).

10 Smithson's readiness to defy theory in the process of doing earth art resonates with scientific calls for open-minded approaches to our twenty-first-century predicaments. For example, after reviewing extensive literature on climate change effects in the Canadian prairies, Sauchyn et al. (2020, 54) conclude: "Incremental adjustments made to historical practices and policies are unlikely to deliver adequate adaptation to the evolving hydroclimate and could potentially be maladaptive over time. Therefore, a significant emerging issue is the need for transformational adaptation that challenges existing policies, structures and systems." See also "Building a Climate Resilient City: Transformational Adaptation" in the *Climate Atlas of the Prairie*, Climate Centre of Canada (4–5). http://prairieclimatecentre.ca/wp-content/uploads/2017/04/pcc-brief-climate-resilient-city-transformational-adaptation.pdf.

11 Mitigating the problem of icy chokepoints (e.g., see note 8.8), Manitobans have invented new machinic routines. For example, a troupe of Wolverine machines weaken a solid ice sheet by scoring it into quadrilaterals. The Wolverines are followed by a troupe of Amphibexes (hybrid amphibious dredgers). Equipped with balancing arms to prevent sinking, the Amphibexes repeatedly throw their heavy bodies on to the scored ice, making holes. Together these small-scale icebreakers, some manned, some not, but all equipped with communication technologies, help meltwaters to flow through the channels rather than bursting out over the banks. The opening of the Amphibex season also provides an opportunity for (meso)political engineering performance. (Steve Topping interview and site visit to North Red Community Water Maintenance Corporation in Selkirk, with Darryl Kupchik, executive director, 25 November 2014). For an Amphibex in action, see https://www.youtube.com/watch?v=zEaNP6F5Jds.

12 I added this last point to the list; the rest are based directly on MIT 2013b, 20–1.
13 In his analysis of the flood of 2011, Scott Forbes, biologist and Lake Manitoba cottager (see chapter 5, part IIA and note 6), told me that rather than diverting the Assiniboine through the Portage Diversion into Lake Manitoba, if MIT had allowed *all* the water through to Winnipeg, the city would have been okay but the Assiniboine might have changed its route (in other words, avulsed).

In fact, avulsion is not especially rare in history and prehistory. A peek into the remote sensing studies of avulsion in floods elsewhere suggests that MIT's reliance on Phase 1 dike fortification to stabilize the Assiniboine may not be enough or may even be counter-productive. In their remote sensing study of avulsions in the "lethal" 2010 Indus River flood in Pakistan and in the 2008 Kosi River flood in India, Syvitski and Brakenridge (2013) conclude that stabilizing individual levees may not enhance flood protection overall and may even worsen risk for those downstream. They advise that "planning for temporary channel diversions to spill both water and sediment during floods is necessary." MIT's use of the singular controlled breach at Hoop and Holler to prevent an uncontrolled breach in the Assiniboine's primary dike system is one example of a temporary channel diversion in use in Manitoba (see chapters 5 and 7). The consequences of avulsion in Pakistan and India were as lethal as they were because of the concentration of populations in the floodplains. In contrast, because most of the array of Assiniboine paleochannels cut through sparsely populated prairie, the calculus of risk, consequence, and intervention would be quite different. Perhaps the sense of security that sparse population allows leads MIT to continue to design and build modern monumental infrastructure, albeit now with substantial consultation with local and Indigenous peoples. See current plans for the Lake Manitoba and Lake St Martin Channel Outlet Project: https://www.gov.mb.ca/mit/wms/lmblsmoutlets/overview/index.html.
14 See notes 3.29 on meshworks and 8.19 on the continuum.
15 Surprisingly, the challenges that climate change poses to scientists cross over into opposing elemental spheres. Doing ethnography about fires and how they move, Petryna (2018, 575) finds that "the mutability of runaway change does not permit a predictive horizon" and that "wildfire intensifies and defies representation, posing operational challenges and appealing for novel engagements with the very concept of projec-

tion." This sounds a great deal like the problem avulsion poses to flood control engineers, despite the radically different geophysical dynamics at play.

16 See, e.g., Dutch Room for the River Programme: https://www.dutchwatersector.com/news/room-for-the-river-programme. Klaver's (2018) geo-philosophy of the river meander highlights the impulse toward multiplicity, focusing in part on the Los Angeles River. Zilberg (2017), in her ongoing ethnographic study of the infrastructural politics of the LA River's revitalization, shows how environmentalists' impulse to restore the floodplain clashes with, among other things, a racial politics materialized as an internal border, a fractal river-water body that replicates the racial antagonisms of the US-Mexico border. For a town planner's take on "Too much water in the city," see White (2010, 41–63). There is also a movement to uncover hidden rivers meandering under hardscaped city neighbourhoods, e.g., see Ann Whiston Spirn's short film about the West Philadelphia Landscape Project, entitled *Buried River* (https://vimeo.com/105794704).

17 Letting a river meander and restoring its riparian forest takes human discipline and cooperation, as Winnipeg city naturalist Kristin Tuchscherer explains to me during a site visit to the Assiniboine Restoration Project (in a remnant of the great forest that is now Assiniboine Park). The assumption is that the river is given room to meander over time while the city remains in place. After extreme floods kill trees, the old river bottom reasserts itself as forest (see note 1.44). But this can only happen if houses do not replace the upland trees and if people build houses at least 100 metres from the banks. It is also crucial that homeowners let duller-looking native plants regrow rather than planting exotics they may find more aesthetically pleasing. (Interview and site visit, 2 and 9 September 2014.)

18 See, for example, Whyte's (2018, 228–9) discussion of an Anishinaabe perspective on "*spiraling temporality* (sense of time) in which it makes sense to consider ourselves as living alongside future and past relatives [or future and past rivers?] simultaneously as we walk through life" (my italics and insertion in brackets). See also Martin et al. 2017. And too, in *A World of Many Worlds*, inspired by the Zapatistas, de la Cadena and Blaser (2018, 4) conceptualize the "pluriverse" as "heterogeneous worldings coming together as a political ecology of practices, negotiating their difficult being together in heterogeneity." The plurilinear possibilities of alliances between the Assiniboine River and Winnipeg, I suggest,

lend geomorphic shape and cultural orientation to an emergent Manitoban pluriverse. See also Escobar 2018.

19 Useful here, too, is Philippopoulos-Mihalopoulos's (2015, 9) concept of the continuum: "the continuum is one yet several, and often overlapping ... a singularity and plurality at the same time ... the continuum fractalizes into a flat surface of multiple lawscapes, various cities, homes, rooms, jungles, planets, all repeating the mechanism of in/visibilisation." It is in and from the continuum that bodies both assemble and withdraw, making spatial justice possible.

20 See Howe's (2019) *Ecologics* for an ethnographic exploration of this question in the context of international wind development in Oaxaca.

21 A Lidaresque eye might reveal a triple-phase ethnographic practice: Phase 1: building up knowledge along familiar conceptual pathways only to tear away; Phase 2: scouting for new conceptual pathways co-produced by rivers and human and more-than-human inhabitants; and Phase 3: letting analysis and writing flow into an array of thematic pathways, some ghostly.

22 Rick Gamble, innovative mayor of Dunnattor on Lake Winnipeg, told me that the whitefish industry is waning in the lake's north and waxing in the lake's south because of the movement of fish with climate change. Site visit and interview, 21 October 2014. Similarly in Scandinavia, see Stammler-Gossman 2012. Climate change impacts on fisheries intensify in Manitoban Cree and Anishinaabe communities impacted by Manitoba Hydro; see also note 8.26.

23 Terms emerging from this ethnography are the sphere of unintended agencies, blended habitus, infrastructural outsiders, internal frontier, political engineering, and the shape of uncertainty. Barry's term technozone (see Appendix II), Philippopoulos-Mihalopoulos's (2015) term spatial justice, and Massey's term throwntogetherness are also important in the geo-cultural frame shift of this book.

24 In a transdisciplinary experiment in collective modelling in the flood-prone market town of Pickering in the interior north of England, Whatmore (2013, 40) finds that "in the context of flooding, knowledge controversies often centre on discrepancies between the first-hand experience of flood events and vernacular knowledge accumulated in affected localities, and the flood science that informs 'evidence-based' flood risk management." The project's hybrid model, which redistributed expert-inhabitant riverine expertise, led to an upstream storage intervention – constructed of small but obstructive structures or bunds

(the "bund model"). At first dismissed as "unviable" by Environment Agency consultants, the model ultimately succeeded when subjected to "public scrutiny sufficient to 'slow down' expert reasoning" (41, 44–5). This kind of hybrid experiment in "political inventiveness" (46), in which local inhabitant knowledge of troubling overland flow patterns informed the engineering of small scale, flexible interventions, could work well in Manitoba and elsewhere. See also notes 3.3 and 5.24.

25 See Archive and Anthropocene Dialogue, Appendix I.
26 Although for the most part, outside the scope of this ethnography, hydroelectric dams in Indigenous territories, such as those built and operated by Manitoba Hydro, have an enormous impact and influence throughout Canada. The Canadian Net-Zero Emissions Accountability Act, Bill C-12, put before the Senate of Canada on 29 June 2021, asks the government to review and account for the negative climate impacts of hydroelectricity projects (including flood-induced greenhouse gas emissions and deforestation) and to institute independent oversight bodies with strong Indigenous representation. (See socio-ecological detail in the Wa Ni Ska Tan speech delivered by Senator McCallum, https://hydroimpacted.ca/canadian-net-zero-emissions-accountability-act-speech.) In comparison, flood control, the focus of this book, involves smaller-scale technozones that tend to be less politically and economically consequential than those of hydroelectricity. Might flood control become a more neutral experimental arena in which to start shifting water governance toward plural modes of spatial justice?
27 These could be inspired by existing Manitoban forms and practices, e.g., large sandbag walls, Amphibex ice-cutters, the Z-dike, mounds, and ring dikes, or by looking to other flood-prone cities around the world – for example, the inflatable rubber dams and spillway gates used in Singapore (see http://www.dyrhoff.co.uk/?post_type=portfolio&p=826) or urban waterfront designs like Hamburg's floating sidewalks, which if adopted in The Forks, for example, could lower the perceived need to keep so much water out of the city. See https://www.hafencity.com/en/urban-development/urban-spaces.

APPENDIX I

1 At twentieth century's end at the University of Michigan, diatoms and their freshwater ecologies inspired ecologist Eugene Stoermer to invent the term Anthropocene for the unforgiving way that humans altered the freshwater stages of the diatom-ecologist encounter and, by extension,

Notes to pages 199–202

ecologies everywhere (2012, cited in Haraway 2016, 44). In collaboration with Dutch atmospheric scientist Paul Crutzen, the Anthropocene concept went global (Crutzen and Stoermer 2000). See also note 6.8.

2 It has clearly become a cultural touchstone and trigger for debate. See, e.g., Yusoff (2013; 2016; 2019), Haraway (2015; 2016), and Schneiderman (2015).

3 Early survey map fragments and historical descriptions of settlement patterns can be found in the Manitoba Archives and Winnipeg Library (e.g., Warkentin 1961). For method, see note 1.58.

4 Steffen et al. 2011, cited in Syvitski et al. (2020, 4).

5 Grill et al. 2019, cited in Syvitski et al. (2020, 5). See also Williams et al. 2014.

6 https://www.un.org/development/desa/indigenouspeoples/declaration-on-the-rights-of-indigenous-peoples.html.

7 http://trc.ca/assets/pdf/Calls_to_Action_English2.pdf.

8 Tomiak et al. (2019, 1).

9 Razack 2016; https://en.wikipedia.org/wiki/Death_of_Tina_Fontaine.

10 https://hydroimpacted.ca/.

11 https://www.gov.mb.ca/mit/wms/floodcontrol/major/historic.html; https://www.gov.mb.ca/flooding/history/index.html.

12 Ibid.

13 Tomiak et al. (2019, 1).

14 Liboiron (2021, 32, note 128).

15 See Gill (2000, 134) for analysis of Manitoba Hydro's "lines of power."

16 Steffen et al. 2011, cited in Syvitski et al. (2020, 3).

17 Syvitski and Kettner 2011, cited in Syvitski et al. (2020, 3).

18 Bumsted (1997, 45–91).

19 Bumsted (1997, 37–43).

20 https://winnipeg.ca/waterandwaste/sewage/systemOperation.stm.

21 Stunden Bower (2011, 117–18).

22 Bumsted (1997, 36).

23 Tomiak et al. (2019, 1).

24 Hallowell (1991, 31).

25 https://heritage.enggeomb.ca/index.php/Winnipeg_Water_Supplies.

26 Bumsted (1997, 26–9).

27 Stunden Bower (2011, 108).

28 MIT's (2013b, 53) account gives no indication that the Fairfield River canal was constructed in Anishinaabe territory:

> For example, following flooding and high water in the late nineteenth century, in the late 1890s the federal Department of Public

Works constructed a 200 feet (61.0 m) wide and 1500 feet (457 m) long canal between the lake and the Fairford River for improving the outflow from the lake and *providing flood relief to settlers* around the lake. It is generally accepted that this early project failed to provide the intended flood relief (Library and Archives Canada, Arthur Meighen Fonds, MG26-1, Vol. 43, Microfilm Reel C-3453, Dunn, T.H., 1915). [My italics]

29 Photo display, Amphibex Operations Base, North Red Waterway Maintenance Corporation, near Selkirk. (Site visit with Steve Topping and Darryl Kupchick, 25 November 2014.)
30 See chapter 2.
31 Bumsted (1997, 20–2).
32 Hogue (2015, 12). [Note, Hogue leaves the accent off Metis.]
33 Bumsted (1997, 8–10).
34 See Coulthard (2014, 83–91) for a historical review of how this and other laws shape the terminology used to refer to first peoples in Canada. See also Liboiron (2021, 32 note 128).
35 Kelsey (2003, 702–23), cited in Smith (2013, 119).
36 Map from Manitoba Archives entitled "Plan of River Lots in the Parishes of St. James and St. Boniface, Province of Manitoba. Urban Library, University of Winnipeg.
37 Hallowell (1991, 31).
38 Hogue (2015, 1–4, 228).
39 A nostalgic caption written for a 1922 fiftieth-anniversary volume of the *Manitoba Free Press* laments that the pastoral scene would disappear within five years of this rendering (reprinted in Huck 2003).
40 Aimée Craft (2014, 5, note 16) finds that "the Anishinaabe names of all of the chiefs who are listed on the written version of Treaty One have bird names. It is possible that the Anishinaabe treaty negotiators may have been of the bird clan and that their role in the negotiations related to the bird clan's responsibilities of leadership and ability to speak on behalf of the group." This suggests that their standing as representatives is analogous to those of the Crown.
41 Hogue (2015, 12, 13).
42 Ibid., 230.
43 Stunden Bower (2011, 22–3).
44 Steinberg and Kristoffersen (2017, 6).
45 Bumsted (1997, 22–3).
46 Ibid., 20–2.

47 Coulthard (2014, 84).
48 Zeller (2000, 87).
49 Bumsted (1997, 16–20).
50 Zeller (2000, 86–7).
51 Pritchett (1942, 36–9); Carter 1968.
52 Ibid.
53 Pritchett (1942, 36).
54 Bumsted (1997, 15).
55 HBC Heritage. "The Royal Charter of the Hudson's Bay Company." https://web.archive.org/web/20151007100328/http://hbcheritage.ca/hbcheritage/collections/archival/charter/charter.
56 Byrne (2016, 93–4).
57 Kives and Scott (2013, 26).
58 Appendix I is a counter-narrative to the Wall Through Time.

APPENDIX II

1 For example, University of Winnipeg biologist Eva Pip (1992, 121) finds that hydroelectric dams on the Nelson River have an impoverishing effect on snail ecologies. This is due in part to bank excavation and the manipulation of water in tributaries, flooding some, drying others. Moreover, she writes: "environmental impact statements seldom address components of the aquatic ecosystem that are not economically important. Thus local changes such as disappearance of species are difficult to document." (Interview 14 November 2014.) This is an example of what Stengers (2009, 3) would call an ethoecological problem of the meso(political): the micro milieu of the snails and the macro milieu of the dams are materially connected yet, nevertheless, fall through the cracks of governance.

   As distinctive energy-using installations that span taxonomic and material space, technozones can also contribute to emerging studies of *technospheres*, especially in their geographical meso-scale of rural-urban interlinkages (cf. Zalasiewicz et al. 2017, 11–16). Compared to the conceptually more bounded but more commonly used terms "system" or "ecology" (as in "media ecology" or "habitat"), Barry's technozone concept makes room for more open, entangled, and invisibilized states-of-being and becoming.

2 There is a piece of transboundary folklore about financing construction of the floodway around Winnipeg, also known as "Duff's Ditch" after its

advocate, Duff Roblin, Manitoba's premier, 1958–67. In the story, Roblin and Canada's prime minister are negotiating the federal contribution to the budget. The conversation is said to have gone something like this:
Canada's Prime Minister: Why should the federal government pay for a project of provincial concern?
Manitoban Premier: If you don't, I'll just put a dam at the border with the United States. Then it will be an international issue!
A federal-provincial agreement followed soon thereafter. (Recounted by Dave Morgan, engineer. Interview 3 October 2014.) In the end, the federal government covered 60 per cent of the cost of excavating 100 million cubic yards of earth over a length of 29.4 miles, more than had been moved for the St Lawrence Seaway, the Panama Canal, or the Suez Canal (Bumsted 1997, 93; see also http://www.mhs.mb.ca/docs/mb_history/42/duffsditch.shtml).
3 Wildlands is a term used in the anthropogenic biome or "anthrome" classification system (Ellis and Ramankutty 2008). Making meso-scale technozones visible reveals that wildlands are far from wild.
4 Harms resulting from paradoxical consequences of an abstract ideal have no solution. Thus, spatial justice for infrastructural outsiders can only be achieved through situated negotiation (cf. Philippopoulos-Mihalopoulos 2015, 35, 40, 178–9).
5 See note 1.85.

APPENDIX III

1 http://www.winnipegfreepress.com/local/265829831.html. For more background on this flood, see https://en.wikipedia.org/wiki/2014_Assiniboine_River_flood.
2 https://en.wikipedia.org/wiki/Humidex.
3 See 2022 design for the Lake Manitoba and Lake St Martin outlet channel project proposal: https://www.gov.mb.ca/mit/wms/lmblsmoutlets/overview/index.html#design.

References

Abram, David. 2010. *Becoming Animal: An Earthly Cosmology.* New York: Pantheon Books.

Adam, Barbara. 1998. *Timescapes of Modernity: The Environment and Invisible Hazards.* New York: Routledge.

Adey, Peter. 2014. *Air: Nature and Culture.* London: Reaktion Books.

Alaimo, Stacy, and Susan Hekman, eds. 2008. *Material Feminisms.* Bloomington: Indiana University Press.

Albris, Kristoffer. 2022. "Where Floods Are Allowed: Climate Adaptation as Defiant Acceptance in the Elbe River Valley." In *Cooling Down: Local Responses to Global Climate Change,* edited by Susanna M. Hoffman, Thomas Hylland Eriksen, and Paulo Mendes, 249–68. New York: Berghahn. https://www.berghahnbooks.com/title/EriksenCooling.

Allen, Andres E., Christopher L. Dupont, Miroslav Oborník, Aleš Horák, Adriano Nunes-Nesi, John P. McCrow, Chris Bowler, et al. 2011. "Evolution and Metabolic Significance of the Urea Cycle in Photosynthetic Diatoms." *Nature* 473: 203–7.

Anand, Nikhil. 2017. *Hydraulic City: Water and the Infrastructures of Citizenship in Mumbai.* Durham, NC: Duke University Press.

Aporta, Claudio. 2009. "The Trail as Home: Inuit and their Pan-Arctic Network of Routes." *Human Ecology* 37, no. 2: 131–46.

Aporta, Claudio, Stephanie C. Kane, and Aldo Chircop. 2018. "Shipping Corridors through the Inuit Homeland." Special issue on "Chokepoints," edited by Ashley Carse, Jason Cons, and Andrew Lakoff in *Limn* 10 (April). https://limn.it/articles/shipping-corridors-through-the-inuit-homeland/.

Aradau, Claudia, and Rens van Munster. 2011. *Politics of Catastrophe: Genealogies of the Unknown.* New York: Routledge.

Bakhtin, Mikhail M. 1981. "Forms of Time and of the Chronotope in the Novel." In *The Dialogic Imagination*, edited by Michael Holquist, translated by Caryl Emerson and Michael Holquist, 84–258. Austin: University of Texas Press.

Ballard, Myrle. 2012. "Flooding Sustainable Livelihoods of the Lake St. Martin First Nation: The Need to Enhance the Role of Gender and Language in Anishinaabe Knowledge Systems." Unpublished doctoral thesis, Natural Resource Institute, University of Manitoba.

Ballard, Myrle, Donna E. Martin, and Shirley Thompson. 2016. *Wounded Spirit: Forced Evacuation of Little Saskatchewan First Nation Elders*. 22 January. https://www.youtube.com/watch?v=PQTubc1LIjY.

Ballard, Myrle, Shirley Thompson, and R. Klatt, producers. 2012. *Flooding Hope: The Lake St. Martin First Nations Story*. 10 October. http://www.youtube.com/watch?v=SQStePF5jeg.

Ballestero, Andrea. 2019. *A Future History of Water*. Durham, NC: Duke University Press.

Barad, Karen. 2010. "Quantum Entanglements and Hauntological Relations of Inheritance: Dis/continuities, Spacetime Enfoldings, and Justice-to-Come." *Derrida Today* 3: 240–68.

– 2012. "'Matter Feels, Converses, Suffers, Desires, Yearns and Remembers': Interview with Karen Barad." In *New Materialism: Interviews and Cartographies*, edited by Rick Dolphijn and Iris van der Tuin, 48–70. Ann Arbor, MI: Open Humanities Press. https://library.oapen.org/bitstream/handle/20.500.12657/33904/Dolphijn-van-der-Tuin_2013_New-Materialism.pdf?sequence=1.

Barry, Andrew. 2006. "Technological Zones." *European Journal of Social Theory* 9: 239–53.

– 2013. *Material Politics: Disputes along the Pipeline*. Hoboken, NJ: John Wiley and Sons.

Barry, John M. 1997. *Rising Tide: The Great Mississippi Flood of 1927 and How It Changed America*. New York: Simon and Schuster.

Basso, Keith H. 1996. *Wisdom Sets in Places: Landscape and Language among the Western Apache*. Albuquerque: University of New Mexico Press.

Bater, Christopher W. 2002. "Meander Migration Rates and Age of the Lower Assiniboine River." In *Prairie Perspectives: Geographical Essays*. Winnipeg: Prairie Division of the Canadian Association of Geographers (PCAG).

Baudrillard, Jean. 1983. *Simulations*. New York: Semiotext(e).

Bennett, Jane. 2004. "The Force of Things: Steps toward an Ecology of Matter." *Political Theory* 32, no. 3: 347–72.

– 2010. *Vibrant Matter: A Political Ecology of Things*. Durham, NC: Duke University Press.

Bennett, Jane, and William Chaloupka. 1993. "Introduction." In *In the Nature of Things: Language, Politics and the Environment*, edited by Jane Bennett and William Chaloupka, vi–xvi. Minneapolis: University of Minnesota Press.

Bennett, Tony, and Patrick Joyce, eds. 2010. *Material Powers: Cultural Studies, History and the Material Turn*. London: Routledge.

Bennion, Helen, Carol D. Sayer, John Tibby, and Hunter J. Carrick. 2010. "Diatoms as Indicators of Environmental Change in Shallow Lakes." In *The Diatoms: Applications for the Environmental and Earth Sciences*, 2nd ed., edited by John P. Smol and Eugene F. Stoermer, 152–73. New York: Cambridge University Press.

Berland, Jody. 2009. *North of Empire. Essays on the Cultural Technologies of Space*. Durham, NC: Duke University Press.

Bierman, Paul R., and David R. Montgomery. 2014. *Key Concepts in Geomorphology*. New York: W.H. Freeman and Company Publishers.

Bjornerud, Marcia. 2018. *Timefulness: How Thinking Like a Geologist Can Help Save the World*. Princeton, NJ: Princeton University Press.

Blais, Eric-Lorne, Jeremy Greshuk, and Tricia Stadnyk. 2016. "The 2011 Flood Event in the Assiniboine River Basin: Causes, Assessments and Damages." *Canadian Water Resources Journal/Revue canadienne des resources hydriques* 41: 74–84.

Blanc, Jacob. 2019. *Before the Flood: The Itaipu Dam and the Visibility of Rural Brazil*. Durham, NC: Duke University Press.

Blunt, Wilfrid. 1971. *The Compleat Naturalist: A Life of Linnaeus*. New York: Viking Press.

Bobbette, Adam, and Amy Donovan. 2019. "Political Geology: An Introduction." In *Political Geology: Active Stratigraphies and the Making of Life*, edited by Adam Bobbette and Amy Donovan, 1–36. New York: Palgrave MacMillan.

Bollas, Christopher. 2008. *The Evocative Object World*. New York: Routledge.

Borrows, John (Kegedonce). 2001. "Indian Agency: Forming First Nations Law in Canada." *PoLAR: Political and Legal Anthropology Review* 24, no. 2: 9–24.

– ed. 2019. *Braiding Legal Orders: Implementing the UN Declaration on the Rights of Indigenous People*. Toronto: Centre for International Governance Innovation.

Bourdieu, Pierre. 1985. *Outline of a Theory of Practice*. Cambridge: Cambridge University Press.

Braun, Bruce. 2000. "Producing Vertical Territory: Geology and Governmentality in Late Victorian Canada." *Ecumene* 7, no. 1: 7–45.

Brondízio, Eduardo S., and Rinku Roy Chowdhury. 2013. "Human-Environment Research: Past Trends, Current Challenges, and Future Directions." In *Human-Environment Interactions: Current and Future Directions*, edited by Eduardo S. Brondízio and Emilio F. Moran, 391–400. Dordrecht: Springer.

Brooks, Gregory R., and Erik Nielsen. 2000. "Red River, Red River Valley, Manitoba." *The Canadian Geographer* 44: 304–9.

Bumsted, John Michael. 1997. *Floods of the Centuries: A History of Flood Disasters in the Red River Valley 1776–1997*. Winnipeg: Great Plains Publications.

Butler, Judith. 2016. *Frames of War: When Is Life Grievable?* New York: Verso.

Byrd, Jodi A. 2011. *The Transit of Empire: Indigenous Critiques of Colonialism*. Minneapolis: University of Minnesota Press.

Byrne, Angela. 2016. "Scientific Practice and the Scientific Self in Rupert's Land, c. 1770–1830: Fur Trade Networks of Knowledge Exchange." In *Spaces of Global Knowledge: Exhibition, Encounter and Exchange in an Age of Empire*, edited by Diarmid A. Finnegan and Jonathan J. Wright, 79–95. New York: Routledge.

Cameron, Emilie. 2017. "Climate Anti-Politics: Scale, Locality, and Arctic Climate Change." In *Ice Blink: Navigating Northern Environmental History*, edited by Stephen Bocking and Brad Marten, 465–95. Calgary: University of Calgary Press.

Cameron, Emilie, Sarah de Leeuw, and Caroline Desbiens. 2014. "Indigeneity and Ontology." *Cultural Geographies* 21, no 1: 19–26.

Canadian Water Security Initiative. 2019. "Water Security for Canadians: Solutions for Canada's Emerging Water Crisis." https://gwf.usask.ca/documents/meetings/water-security-for-canada/WaterSecurityForCanada_April-25-2019-2pg1.pdf.

Caretta, Martina Angela, Aditi Mukherji, Md Arfanuzzaman, Richard A. Betts, Alexander Gelfan, Yukiko Hirabayashi, Tabea Katharina Lissner ... and Seree Supratid. 2022. "Water." In *Climate Change 2022: Impacts, Adaptation, and Vulnerability*. Contribution of Working Group II to the Sixth Assessment Report of the Intergovernmental Panel on Climate Change, edited by Hans-O. Pörtner, Debra C. Roberts, M. Tignor, Elvira Poloczanska, Katja Mintenbeck, A. Alegría ... and M. Craig, 4.1–4.413. Cambridge University Press. In Press.

Carse, Ashley. 2014. *Beyond the Big Ditch: Politics, Ecology, and Infrastructure at the Panama Canal*. Cambridge, MA: MIT Press.

Carter, George E. 1968. "Lord Selkirk and the Red River Colony." *Montana: The Magazine of Western History* 18, no. 1: 60–9.

Clarke, Garry K.C., David W. Leverington, James T. Teller, and Arthur S. Dyke. 2004. "Paleohydraulics of the Last Outburst Flood from Glacial Lake Agassiz and the 8200 BP Cold Event." *Quarternary Science Reviews* 23: 389–407.

Clifford, James. 2013. *Returns: Becoming Indigenous in the Twenty-First Century*. Cambridge, MA: Harvard University Press.

Clifford, James, and George Marcus, eds. 1986. *Writing Culture: The Poetics and Politics of Ethnography*. Berkeley: University of California Press.

Collier, Stephen J., James C. Mizes, and Antina von Schnitzler, eds. "Public Infrastructures / Infrastructural Publics." *Limn* 7 (July). https://limn.it/issues/public-infrastructuresinfrastructural-publics/.

Consumers Association of Canada (Manitoba Branch). 2014. "Keeyask – A Watershed Decision. Closing Arguments Submitted to the CEC." 14 January. Prepared by Byron Williams, Aimée Craft, and Joëlle Pastora Sala. http://www.cecmanitoba.ca/cecm/hearings/pubs/Keeyask_Generation_Project/Presentations/CAC_CLOSING_ARGUMENTS_-_JAN_14_2014.pdf.

Coulthard, Glen Sean. 2014. *Red Skin White Masks: Rejecting the Colonial Politics of Recognition*. Minneapolis: University of Minnesota Press.

Craft, Aimée. 2014. "Living Treaties, Breathing Research." *Canadian Journal of Women and the Law* 26, no.1: 1–22.

Craft, Aimée, and Paulette Regan. 2020a. "Introduction." In *Pathways of Reconciliation: Indigenous and Settler Approaches to Implementing the TRC's Calls to Action*, edited by Aimée Craft and Paulette Regan, xi–xxi. Winnipeg: University of Manitoba Press.

– eds. 2020b. *Pathways of Reconciliation: Indigenous and Settler Approaches to Implementing the TRC's Calls to Action*. Winnipeg: University of Manitoba Press.

Cruikshank, Julie. 2005. *Do Glaciers Listen? Local Knowledge, Colonial Encounters, and Social Imagination*. Vancouver: University of British Columbia Press.

Crutzen, Paul J., and Eugene F. Stoermer. 2000. "The 'Anthropocene.'" *IGBP Newsletter* 41: 17–18.

Currie, Buzz, and *Winnipeg Free Press* staff. 1997. *A Red Sea Rising: The Flood of the Century*. Winnipeg: *Winnipeg Free Press*.

Da Cunha, Dilip. 2019. *The Invention of Rivers: Alexander's Eye and Ganga's Descent*. Philadelphia: University of Pennsylvania Press.

Daigle, Michelle. 2019. "The Spectacle of Reconciliation: On (the) Unsettling Responsibilities to Indigenous Peoples in the Academy." *EPD: Society and Space* 37, no. 4: 703–21.

De la Cadena, Marisol, and Mario Blaser, eds. 2018. *A World of Many Worlds*. Durham NC: Duke University Press.

De Landa, Manuel. 1997. *A Thousand Years of Non-Linear History*. New York: Zone Books.

Derrida, Jacques, and Bernard Stiegler. 2002. *Echographies of Television: Filmed Interviews*. Cambridge: Polity.

De Wolff, Kim, Rina C. Faletti, and Ignacio López-Calvo, eds. 2022. *Hydrohumanities: Water Discourse and Environmental Futures*. Berkeley: University of California. Open access. https://luminosoa.org/site/books/e/10.1525/luminos.115/.

Dodds, Klaus, and Mark Nuttall. 2016. *The Scramble for the Poles*. Cambridge: Polity.

Dundes, Alan, ed. 1988. *The Flood Myth*. Berkeley: University of California Press.

Dyce, Matthew. 2013. "Canada between the Photograph and the Map: Aerial Photography, Geographical Vision and the State." *Journal of Historical Geography* 39: 69–84.

Edgeworth, Matt. 2011. *Fluid Pasts: Archaeology of Flow*. London: Bristol Classical Press.

Elden, Stuart. 2010. "Land, Terrain, Territory." *Progress in Human Geography* 34: 799–817.

– 2019. "The Instability of Terrain: On Limits and Boundaries." In *A Moving Border: Alpine Cartographies of Climate Change*, edited by Marco Ferrari, Elisa Pasqual, and Andrea Bagnato, 51–61. ZKM | Center for Arts and Media and New York: Columbia University Press.

Ellis, Erle C., and Navin Ramankutty. 2008. "Putting People in the Map: Anthropogenic Biomes of the World." *Frontiers in Ecology and the Environment* 6: 439–47.

Elson, John A. 1967. "Geology of Glacial Lake Agassiz." In *Life, Land and Water*, Proceedings of 1966 Conference on Environmental Studies of the Glacial Lake Agassiz Region, edited by William J. Mayer-Oakes, 37–95. Winnipeg: University of Manitoba Press.

– 1983. "Lake Agassiz – Discovery and a Century of Research." In *Glacial Lake Agassiz*, edited by James T. Teller and Lee Clayton, 21–42. Toronto: University of Toronto Press.

Elyachar, Julia. 2010. "Phatic Labor, Infrastructure, and the Question of Empowerment in Cairo." *American Ethnologist* 37, no. 3: 452–64.

Escobar, Arturo. 2008. *Territories of Difference: place, movements, life, redes*. Durham, NC: Duke University Press.
— 2018. *Designs for the Pluriverse: Radical Interdependence, Autonomy, and the Making of Worlds*. Durham, NC: Duke University Press.
European Commission (EC). 2011. "Urea Cycle behind Evolutionary Success of Diatoms." https://cordis.europa.eu/news/rcn/33397_en.html.
Evers, Jaap. 2019. "The Time to Influence the Dutch Water Democracy Is NOW (March 20)." *FLOWs*, 19 March. https://flows.hypotheses.org/3437.
Fabian, Johannes. 1983. *Time and the Other: How Anthropology Makes Its Object*. New York: Columbia University Press.
Faulkner, William. [1939]1948. *The Old Man: Violence and Terror in a Mississippi Flood*. New York: Signet Books.
Ferrari, Marco, Elisa Pasqual, and Andrea Bagnato, eds. 2019. *A Moving Border: Alpine Cartographies of Climate Change*. ZKM | Center for Arts and Media and New York: Columbia University Press.
Fujikane, Candace. 2021. *Mapping Abundance for a Planetary Future: Kanaka Maoli and Critical Settler Cartographies in Hawai'i*. Durham, NC: Duke University Press.
Gandy, Matthew. 2002. *Concrete and Clay: Reworking Nature in New York City*. Cambridge, MA: MIT Press.
Gill, Shelia Dawn. 2000. "The Unspeakability of Racism: Mapping Law's Complicity in Manitoba's Racialized Spaces." *Canadian Journal of Law and Society* 15, no. 2: 131–62.
Gober, Patricia, and Howard S. Wheater. 2013. "Socio-hydrology and the Science-Policy Interface: A Case Study of the Saskatchewan River Basin." *Hydrology and Earth System Sciences* 10: 6,669–93.
Gordillo, Gastón. 2021. "The Power of Terrain: The Affective Materiality of Planet Earth in the Age of Revolution." *Dialogues in Human Geography* 11, no. 2: 190–4.
Gould, Stephen Jay. 1981. *The Mismeasure of Man*. New York: W.W. Norton and Company.
Government of Canada. 2019. *Canada's Changing Climate Report*. Ottawa. https://www.nrcan.gc.ca/climate-change/impacts-adaptations/10761.
Graham, Stephen, ed. 2010. *Disrupted Cities: When Infrastructure Fails*. London: Routledge.
Greater Winnipeg Dyking Board. 1951. *Final Report on the Activities of the Greater Winnipeg Dyking Board. To The Right Honourable C.D. Howe, Minister of Trade and Commerce, Ottawa, Canada and The Honourable D.L. Campbell, Premier of Manitoba*, Appendix G. Winnipeg, 1 October.

Gregory, Derek. 1994. *Geographical Imaginations*. Oxford: Blackwell.
Grek Martin, Jason. 2014. "Survey Science on Trial: The Geographic Contours of Geology's Practical Science Debate in Late Victorian Canada." *Journal of Historical Geography* 45: 1–11.
Grill, Günther, Bernhard Lehner, Michele Thieme, B. Greenen, David Tickner, Francesca Antonelli, and Suresh Babu (+ 27 more). 2019. "Mapping the World's Free-Flowing Rivers." *Nature* 569: 215–21.
Guano, Emanuela. 2017. *Creative Urbanity: An Italian Middle Class in the Shade of Revitalization*. Philadelphia: University of Pennsylvania Press.
Håkansson, Hannelore, and Hedy Kling. 1994. "*Cyclotella Agassizensis* Nov. Sp. and Its Relationship with *C. Quillensis* Bailey and Other Prairie *Cyclotella* Species." *Diatom Research* 9, no. 2: 289–301.
Hallowell, A. Irving. 1991. *The Ojibwa of Berens River, Manitoba: Ethnography into History*. Fort Worth: Harcourt Brace Jovanovich College Publishers.
Haque, C. Emdad. 2000. "Risk Assessment, Emergency Preparedness and Response to Hazards: The Case of the 1997 Red River Valley." *Natural Hazards* 21: 225–45.
Haraway, Donna. 1985. "A Manifesto for Cyborgs: Science, Technology, and Socialist Feminism in the 1980s." *Socialist Review* 15, no. 2: 65–108.
– 1988. "Situated Knowledges: The Science Question in Feminism and the Privilege of Partial Perspective." *Feminist Studies* 14, no. 3: 575–99.
– 2015. "Anthropocene, Capitalocene, Plantationocene, Chthulucene: Making Kin." *Environmental Humanities* 6: 159–65.
– 2016. *Staying with the Trouble: Making Kin in the Chthulucene*. Durham, NC: Duke University Press.
Harvey, Penny, and Hannah Knox. 2015. *Roads: An Anthropology of Infrastructure and Expertise*. Ithaca, NY: Cornell University Press.
Helmreich, Stefan. 2003. "A Tale of Three Seas: From Fishing through Aquaculture to Marine Biotechnology in the Life History Narrative of a Marine Biologist." *Maritime Studies* 2, no. 2: 73–94.
Henrici, Jane. 2002. "Speed and Space within a NAFTA Corridor." *Space and Culture* 5: 49–52.
Hetherington, Kregg, ed. 2019. *Infrastructure, Environment, and Life in the Anthropocene*. Durham, NC: Duke University Press.
Heynen, Nik, Maria Kaika, and Erik Swyngedouw, eds. 2006. *In the Nature of Cities: Urban Political Ecology and the Politics of Urban Metabolism*. London: Routledge.
Hoffman, Susanna M. 2020. "The Scope and Importance of Anthropology of Disaster to Practitioner Settings and Policy Creation." In *Disaster upon Disaster: Exploring the Gap between Knowledge, Policy, and Practice*," edited

by Susanna M. Hoffman and Roberto E. Barrios, 269–91. New York: Berghahn.

Hogue, Michel. 2015. *Metis and the Medicine Line: Creating a Border and Dividing a People*. Chapel Hill: University of North Carolina Press.

Hooimeijer, Fransje. 2014. *The Making of Polder Cities: A Fine Dutch Tradition*. Prinsenbeek: Jap Sam Books.

Howe, Cymene. 2019. *Ecologics: Wind and Power in the Anthropocene*. Durham, NC: Duke University Press.

Huck, Barbara. 2003. *Crossroads of the Continent: A History of the Forks and Assiniboine Rivers*. Winnipeg: Heartland Associates, Inc.

Hughes, Francesca. 2019. "Inequalities of Ice: Francesca Hughes on the Seductions of Measure." In *A Moving Border: Alpine Cartographies of Climate Change*, edited by Marco Ferrari, Elisa Pasqual, and Andrea Bagnato, 129–37. ZKM | Center for Arts and Media and New York: Columbia University Press.

Hustak, Carla, and Natasha Myers. 2012. "Involutionary Momentum. *differences* 23, no. 3: 74–118.

Irmscher, Christoph. 2013. *Louis Agassiz*. Boston: Houghton Mifflin Harcourt.

Jacob, Christian. 2006. *The Sovereign Map: Theoretical Approaches to Cartography throughout History*, translated by Tom Conley. Chicago: University of Chicago Press.

Jameson, Fredric. 1981. *The Political Unconscious*. Ithaca, NY: Cornell University Press.

Jerolmack, Doug. 2014. "Tuning Sediment." In *Design in the Terrain of Water*, edited by Anuradha Mathur and Dilip da Cunha, 75–81. Philadelphia: University of Pennsylvania School of Design.

Jiménez, Alberto Corsín. 2017. "Exchanging Equations: Anthropology as/beyond Symmetry." In *Redescribing Relations: Strathernian Conversations on Ethnography, Knowledge and Politics*, edited by Ashley Lebner, 77–103. New York: Berghahn Books.

Julius, Matthew L., and Edward C. Theriot. 2010. "The Diatoms: A Primer." In *The Diatoms: Applications for Environmental and Earth Sciences*, 2nd ed., edited by John P. Smol and Eugene F. Stoermer, 8–22, New York: Cambridge University Press.

Kane, Stephanie C. 1994. *The Phantom Gringo Boat: Shamanic Discourse and Development in Panama*. Washington, DC: Smithsonian Institution Press. 1st ed., IU ScholarWorks: https://scholarworks.iu.edu/dspace/handle/2022/22251.

– 2010. "Animated Architecture: Maria Luiza Mendez Lins and the Colonial Water Taps of Olinda, Brazil." *Journal of Folklore Research* 46, no. 3: 293–324.

— 2011. "Visibility and Contamination on the Buenos Aires Waterfront: Under the Bridges of Puerto Madero and La Boca." In *Transforming Urban Waterfronts: Fixity and Flow*, edited by Gene Desfor, Jennefer Laidley, Quentin Stevens, and Dirk Schubert, 211–34, New York: Routledge.

— 2012. *Where Rivers Meet the Sea: The Political Ecology of Water*. Philadelphia: Temple University Press.

— 2016. "Reestablishing the Fundamental Bases for Environmental Health: Infrastructure and the Socio-topographies of Surviving Seismic Disaster." In *A Companion to Environmental Health: Anthropological Perspectives*, edited by Merrill Singer, 348–72. Malden, MA: John Wiley and Sons.

— 2017. "Enclave Ecology: Hardening the Land-Sea Edge to Provide Freshwater in Singapore's Hydrohub." *Human Organization* 76, no. 1: 82–95.

— 2018a. "Engineering an Island City-State: A 3D Ethnographic Comparison of the Singapore River and Orchard Road." In *Rivers of the Anthropocene*, edited by Jason M. Kelly, Philip Scarpino, Helen Berry, J.P.M. Syvitski, and Michel Meybeck, 135–49. Berkeley: University of California Press. Open access.

— 2018b. "Where Sheets of Water Intersect: Infrastructural Logistics and Sensibilities in Winnipeg, Manitoba." In *Territory beyond Terra*, edited by Kimberley Peters, Philip Steinberg, and Elaine Stratford, 107–26, London: Rowman and Littlefield.

— 2019. "Circumpolarity and the Trophic Architecture of Ice: Animals in Ecology and Law." Paper presented at Ice Law Project conference, Durham University, Durham, UK, 26 April.

— 2022. "Winnipeg's Aspirational Port and the Future of Arctic Shipping (The Geo-Cultural Version)." In *Hydrohumanities: Water Discourse and Environmental Futures*, edited by Kim De Wolff, Rina C. Faletti, and Ignacio López-Calvo, 42–63, Berkeley: University of California Press. Open access. https://doi.org/10.1525/luminos.115.e.

Kane, Stephanie C., Eden Medina, and Daniel M. Michler. 2015. "Infrastructural Drift in Seismic Cities: Chile, Pacific Rim, 27 February 2010." *Social Text* 33, no. 122: 71–92.

Kelly, Jason M., Philip Scarpino, Helen Berry, J.P.M. Syvitski, and Michel Meybeck, eds. 2018. *Rivers of the Anthropocene*. Berkeley: University of California Press. http://library.oapen.org/handle/20.500.12657/31005.

Kelsey, Robin E. 2003. "Viewing the Archive: Timothy O'Sullivan's Photographs for the Wheeler Survey, 1871–4." *The Art Bulletin* 85, no. 4: 702–23.

Kirchmeier-Young, M., and Xuebin Zhang. 2020. "Human Influence Has Intensified Extreme Precipitation in North America." *Proceedings of the National Academy of Sciences*. https://www.pnas.org/lookup/suppl/doi:10.1073/pnas.1921628117/-/DCSupplemental.

Kives, Bartley. 2015. "How's My Home, James? What the Measurement Means to Flood Prone Winnipeg." *Winnipeg Free Press*, 26 March. https://www.winnipegfreepress.com/local/hows-my-home-james-297619021.html.

Kives, Bartley, and Bryan Scott. 2013. *Stuck in the Middle: Dissenting Views of Winnipeg*. Winnipeg: Great Plains Publications.

Klaver, Irene J. 2005. "Stone Worlds: Phenomenology on (the) Rocks." In *Environmental Philosophy: From Animal Rights to Radical Ecology*, 4th ed., edited by Michael E. Zimmerman, Karen J. Warren, Irene Klaver, and John Clark, 347–59. Upper Saddle River, NJ: Pearson Prentice Hall.

– 2017. "Water, Mud, and Sand: Dutch Re-scaping the Land." In *Hypernatural Landscapes in the Anthropocene*, edited by Sabine Flach and Gary Sherman, 101–22. Oxford: Peter Lang International Academic Publishers.

– 2018. "Meandering and Riversphere: The Potential of Paradox." *Open Rivers: Rethinking Water, Place and Community* 11: 45–65. https://editions.lib.umn.edu/openrivers/wp-content/uploads/sites/9/2018/08/openrivers_issue_11_book_sm.pdf.

Knowles, Scott. 2014. "Engineering Risk and Disaster: Disaster-STS and the American History of Technology." *Engineering Studies* 6, no. 3: 227–48.

Kulchyski, Peter. 2005. *Like the Sound of a Drum: Aboriginal Cultural Politics in Denendeh and Nunavut*. Winnipeg: University of Manitoba Press.

Lambert, Steve. 2020. "Dikes, Ditches and Dams: Manitoba's Fight against Flooding Is Complex." *Canada's National Observer*, 3 January. https://www.nationalobserver.com/2020/01/03/news/dikes-ditches-and-dams-manitobas-fight-against-flooding-complex.

Last, William M. 1984. "Modern Sedimentology and Hydrology of Lake Manitoba, Canada." *Environmental Geology* 5: 177–90.

Law, John 1994. *Organizing Modernity*. Oxford: Blackwell.

– 2009. "Actor Network Theory and Material Semiotics." In *The New Blackwell Companion to Social Theory*, edited by Bryan S. Turner, 141–58. Hoboken, NJ: Wiley-Blackwell.

Lebner, Ashley. 2017. "Introduction." In *Redescribing Relations: Strathernian Conversations on Ethnography, Knowledge and Politics*, edited by Ashley Lebner, 1–37. New York: Berghahn Books.

Lefebvre, Henri. 1991. *The Production of Space*, translated by Donald Nicholson-Smith. Malden, MA: Blackwell.

Leon, Alanna Young, and Denise Marie Nadeau. 2017. "Moving with Water: Relationships and Responsibilities." In *Downstream: Reimagining Water*, edited by Dorothy Christian and Rita Wong, 117–38, Waterloo, ON: Wilfrid Laurier University Press.

Li, Lanhai, and Slobodan P. Simonovic. 2002. "System Dynamics Model for Predicting Floods from Snowmelt in North American Prairie Watersheds." *Hydrological Processes* 16: 2,645–66.

Liboiron, Max. 2021. *Pollution Is Colonialism*. Durham, NC: Duke University Press.

Lightfoot, Sheryl. 2016. *Global Indigenous Politics: A Subtle Revolution*. New York: Routledge.

– 2020. "Conclusion." In *Pathways of Reconciliation: Indigenous and Settler Approaches to Implementing the TRC's Calls to Action*, edited by Aimée Craft and Paulette Regan, 280–9. Winnipeg: University of Manitoba Press.

Lindeman, Raymond. 1942. "The Trophic-Dynamic Aspect of Ecology." *Ecology* 23, no. 4: 399–417.

Linton, Jamie. 2010. *What Is Water? The History of a Modern Abstraction*. Vancouver: University of British Columbia Press.

Manitoba Flood Relief Fund. 2000. *River Rampant*. 2nd ed. Steinbach, MB: Derksen Printers Ltd.

Manitoba Hydro. 2011. *Powering the Province: Sixty Years of Manitoba Hydro*. Winnipeg: Manitoba Hydro.

Martin, Donna E., Shirley Thompson, Myrle Ballard, and Janice Linton. 2017. "Two-Eyed Seeing in Research and Its Absence in Policy: Little Saskatchewan First Nation Elders' Experiences of the 2011 Flood and Forced Displacement." *The International Indigenous Policy Journal* 8, no. 4, Art. 6.

Massey, Doreen. 2005. *for space*. Los Angeles: Sage.

McKittrick, Katherine. 2021. *Dear Science and Other Stories*. Durham, NC: Duke University Press.

Million, Dian. 2009. "Felt Theory: An Indigenous Feminist Approach to Affect and History." *Wicazo Sa Review* 24, no. 2: 53–76.

Milly, P. Chris D., Julio Betancourt, Malin Falkenmark, Robert M. Hirsch, Zbigniew Kundzewicz, Dennis P. Lettenmaier, and Ronald J. Stouffer. 2008. "Stationarity Is Dead: Whither Water Management?" *Science* 319, no. 5,863: 573–4.

Mirzoeff, Nicholas. 2011. *The Right to Look: A Counterhistory of Visuality*. Durham, NC: Duke University Press.

MIT (Manitoba Infrastructure and Transport). 2013a. *Red River Floodway*

Operation Report. https://www.gov.mb.ca/mit/floodinfo/floodproofing/reports/pdf/2013_red_river_floodway_operation_report.pdf.

– 2013b. *2011 Flood: Technical Review of Lake Manitoba, Lake St. Martin, and Dauphin River and Related Issues*. Manitoba 2011 Flood Review Task Force Report to the Ministry of Conservation. April. Winnipeg. http://content.gov.mb.ca/mit/wm/assiniboine_lakemb_lsm_report_nov2013.pdf.

Mol, Annemarie. 2002. *The Body Multiple: Ontology in Medical Practice*. Durham, NC: Duke University Press.

Morris, Norval. 1992. *The Brothel Boy and Other Parables in the Law*. Oxford: Oxford University Press.

Mukerji, Chandra. 2009. *Impossible Engineering: Technology and Territoriality on the Canal du Midi*. Princeton, NJ: Princeton University Press.

Murchie, Guy. 1954. *Song of the Sky: An Exploration of the Ocean of Air*. Boston: Houghton Mifflin.

Napoleon, Val. 2007. "Thinking about Indigenous Legal Orders." *Derecho & Sociedad* 42: 137–46.

– 2009. "Ayook: Gitksan Legal Order, Law and Legal Theory." Unpublished doctoral thesis, Faculty of Law, University of Victoria. https://dspace.library.uvic.ca:8443/handle/1828/1392.

Nixon, Rob. 2011. *Slow Violence and the Environmentalism of the Poor*. Cambridge, MA: Harvard University Press.

Odoni, Nicholas A., and Stuart N. Lane. 2010. "Knowledge-Theoretic Models in Hydrology." *Progress in Physical Geography* 34, no. 2: 151–71.

O'Reilly, Jessica, Naomi Oreskes, and Michael Oppenheimer. 2012. "The Rapid Disintegration of Projections: The West Antarctic Ice Sheet and Intergovernmental Panel on Climate Change." *Social Studies of Science* 42, no. 5: 709–31.

Oreskes, Naomi. 2007. "From Scaling to Simulation: Changing Meanings and Ambitions of Models in Geology." In *Science without Laws: Model Systems, Cases, Exemplary Narratives*, edited by Angela N.H. Creager, Elizabeth Lunbeck, and M. Norton Wise, 93–124. Durham, NC: Duke University Press.

Overeem, Irina, and J.P.M. Syvitski. 2010. "Shifting Discharge Peaks in Arctic Rivers, 1997–2007." *Geografiska Annaler* 92, no. 2: 285–96.

Owen, Bruce, and Larry Kusch. 2014. "Province to Breach Hoop and Holler Dike." *Winnipeg Free Press*, 4 July. http://www.winnipegfreepress.com/local/265829831.html.

Pauls, Karen. 2019. "Manitoba Spent Money to Fight Flooding. Could Its Solutions Work in Eastern Canada?" *CBC News*, 1 May. https://www.cbc.ca

/news/canada/manitoba/manitoba-spent-big-money-to-fight-flooding-could-its-solutions-work-in-eastern-canada-1.5117566.

Pearce, Peter H., Francoise Bertrand, and James W. MacLaren. 1985. *Currents of Change: Final Report, Inquiry on Federal Water Policy*. Ottawa: Government of Canada.

Peters, Kimberley, Philip Steinberg, and Elaine Stratford, eds. 2018. *Territory beyond Terra*. London: Rowman and Littlefield.

Petryna, Adriana. 2018. "Wildfires at the Edges of Science: Horizoning Work amid Runaway Change." *Cultural Anthropology* 33, no. 4: 570–95.

Philippopoulos-Mihalopoulos, Andreas. 2015. *Spatial Justice: Body, Lawscape, Atmosphere*. New York: Routledge.

Pip, Eva. 1992. "The Ecology of Subarctic Mollusks in the Lower Nelson River System, Manitoba, Canada." *Journal of Molluscan Studies* 58: 121–6.

Povinelli, Elizabeth A. 2016. *Geontologies: A Requiem to Late Liberalism*. Durham, NC: Duke University Press.

Prairie Climate Centre. 2019. "Winnipeg, Manitoba." In *Climate Atlas of Canada, version 2*. University of Winnipeg. https://climateatlas.ca/sites/default/files/cityreports/Winnipeg-EN.pdf.

Pratt, Mary Louise. 1992. *Imperial Eyes: Travel Literature and Transculturation*. New York: Routledge.

– 2017. "Coda: Concept and Chronotope." In *Arts of Living on a Damaged Planet*, edited by Anna Tsing, Heather Swanson, Elaine Gan, and Nils Bubandt, 169–74. Minneapolis: University of Minnesota Press.

Pritchard, Sara B. 2011. *Confluence: The Nature of Technology and the Remaking of the Rhône*. Cambridge, MA: Harvard University Press.

Pritchett, John Perry. 1942. *The Red River Valley 1811–1849: A Regional Study*. New Haven, CT: Yale University Press.

Rademacher, Anne M. 2011. *Reigning the River: Urban Ecologies and Political Transformation in Kathmandu*. Durham, NC: Duke University Press.

Raffles, Hugh. 2002. *In Amazonia: A Natural History*. Princeton, NJ: Princeton University Press.

Rannie, William F. 1999. "Manitoba Flood Protection and Control Strategies." *Proceedings of the North Dakota Academy of Science* 53: 38–43.

– 2017. "Landscapes of the Assiniboine River Watershed." In *Landscapes and Landforms of Western Canada*, edited by Olav Slaymaker, 131–42. Basel: Springer International.

Rannie, William F., Leonard H. Thorleifson, and James T. Teller. 1989. "Holocene Evolution of the Assiniboine River Paleochannels and Portage la Prairie Alluvial Fan." *Canadian Journal of Earth Sciences* 26: 1834–41.

Rashid, Harun. 2011. "Interpreting Flood Disasters and Flood Hazard Perceptions from Newspaper Discourse: Tale of Two Floods in the Red River Valley, Manitoba, Canada." *Applied Geography* 31: 35–45.

Razack, Sherene H. 2002. "When Place Becomes Race, Introduction." In *Race, Space, and the Law: Unmapping a White Settler Society*, edited by Sherene H. Razack, 1–20. Toronto: Between the Lines.

– 2016. "Sexualized Violence and Colonialism: Reflections on the Inquiry into Missing and Murdered Indigenous Women." *Canadian Journal of Women and the Law* 28, no. 2: i–viii.

Ribeiro, Gustavo Lins. 1994. *Transnational Capitalism and Hydropolitics in Argentina: The Yacyretá High Dam*. Gainesville: University Press of Florida.

Risberg, Jan, Per Sandgren, James T. Teller, and William M. Last. 1999. "Siliceous Microfossils and Mineral Magnetic Characteristics in a Sediment Core from Lake Manitoba, Canada: A Remnant of Glacial Lake Agassiz." *Canadian Journal of Earth Sciences* 36: 1, 299–314.

Ross, Ian. 2001. *The Gap*. Toronto: J. Gordon Shillingford Publishing Inc.

Round, Frank E., Richard M. Crawford, and David G. Mann. 1990. *The Diatoms: Biology and Morphology of the Genera*. Cambridge: Cambridge University Press.

Rudwick, Martin J.S. 2009. "Biblical Flood and Geological Deluge: The Amicable Dissociation of Geology and Genesis." *Geological Society, London, Special Publications* 310, no. 1: 103–10.

Russell, Frances. 2004. *Mistehay Sakahegan, The Great Lake: The Beauty and the Treachery of Lake Winnipeg*. Winnipeg: Heartland Association.

Said, Edward W. 1994. *Culture and Imperialism*. New York: Vintage.

Sandford, Robert W., and Kerry Freek. 2014. *Flood Forecast: Climate Risk and Resiliency in Canada*. Victoria: Rocky Mountain Books.

Sauchyn, David, Debra Davidson, and Mark Johnston. 2020. "The Prairie Provinces." In *Canada in a Changing Climate: Regional Perspectives Report*, edited by Fiona J. Warren, Nicole Lulham, and Donald S. Lemmen, chapter 4. Ottawa. https://changingclimate.ca/regional-perspectives/chapter/4-0/.

Scarry, Elaine. 2012. *Thinking in an Emergency*. New York: Norton.

Schmidt, Jeremy J. 2020. "Settler Geology: Earth's Deep History and the Governance of *in situ* Oil Spills in Alberta. *Political Geography* 78: 102132. https://doi.org/10.1016/j.polgeo.2019.102132.

Schneiderman, Jill S. 2015. "Naming the Anthropocene." *philoSOPHIA* 5, no. 2: 179–201.

Schoch-Spana, Monica, Crystal Franco, Jennifer B. Nuzzo, and Christiana Usenza. 2007. "Community Engagement: Leadership Tool for Catastrophic

Health Events." *Biosecurity and Bioterrorism: Biodefense Strategy, Practice, and Science* 5, no. 1: 8–25.

Schwartz, John. 2019. "As Floods Keep Coming, City Pays Residents to Move." *New York Times*, 6 July. https://www.nytimes.com/2019/07/06/climate/nashville-floods-buybacks.html?action=click&module=Top%20Stories&pgtype=Homepage.

Serres, Michel. 1974. *La Traduction: Hermes III*. Paris: Les Éditions de Minuit.

Shiva, Vandana. 2002. *Water Wars: Privatization, Pollution, and Profit*. Cambridge, MA: South End Press.

Shrubsole, Dan. 2001. "The Cultures of Flood Management in Canada: Insights from the 1997 Red River Experience." *Canadian Water Resources Journal* 26, no. 4: 461–79.

Simms, Rosie, Merrell-Ann Phare, Oliver M. Brandes, and Michael Miltenberger. 2018. "Collaborative Consent as a Path to Realizing UNDRIP." *Policy Options Politiques*, 11 January. https://poliswaterproject.org/files/2018/01/2018-01-11_PolicyOptions_CollaborativeConsent.pdf.

Simonovic, Slobodan P., and Richard W. Carson. 2003. "Flooding in the Red River Basin – Lessons from Post Flood Activities." *Natural Hazards* 28, nos. 2–3: 345–65.

Simpson, Leanne Betasamosake. 2008. "Looking after Gdoo-naaganinaa: Precolonial Nishnaabeg Diplomatic and Treaty Relationships." *Wicazo Sa Review* 23, no. 2: 29–42.

– 2016. "Indigenous Resurgence and Co-resistance." *Critical Ethnic Studies* 2, no. 2: 19–34.

– 2017a. *This Accident of Being Lost*. Canada: House of Anansi Press.

– 2017b. *As We Have Always Done: Indigenous Freedom through Radical Resistance*. Minneapolis: University of Minnesota Press.

Smith, Neil. 1996. *The New Urban Frontier: Gentrification and the Revanchist City*. New York: Routledge.

Smith, Shawn Michelle. 2013. *At the Edge of Sight: Photography and the Unseen*. Durham, NC: Duke University Press.

Smithson, Robert. [1968] 1996. "A Provisional Theory of Non-sites." In *Robert Smithson: The Collected Writings*, edited by Jack Flam, 364–6. Berkeley: University of California Press.

Smol, John P. and Eugene F. Stoermer. 2010. *The Diatoms: Applications for the Environmental and Earth Sciences*, 2nd ed. New York: Cambridge University Press.

Soares, Pedro Paulo de Miranda Araújo. 2022. "Urban Transformations in the Hydric Landscapes of Belém, Brazil." In *Cooling Down: Local Responses*

to *Global Climate Change*, edited by Susanna M. Hoffman, Thomas Hylland Eriksen, and Paulo Mendes, 90–107. New York: Berghahn.
Soll, David. 2013. *Empire of Water: An Environmental and Political History of the New York City Water Supply*. Ithaca, NY: Cornell University Press.
Spinoza, Baruch. 1992. *Ethics: Treatise on the Emendation of the Intellect, and Selected Letters*, translated by Samuel Shirley, edited by Seymour Feldman. Indianapolis: Hackett Publishing.
Stammler-Gossmann, Anna. 2012. "The Big Water of a Small River: Flood Experiences and a Community Agenda for Change." In *Governing the Uncertain: Adaptation and Climate in Russia and Finland*, edited by M. Tenneberg, 55–81. Dordrecht: Springer.
Steffen, Will, Jaques Grinevald, Paul Crutzen, and John McNeill. 2011. "The Anthropocene: Conceptual and Historical Perspectives." *Philosophical Transactions of the Royal Society* A 369: 842–67.
Steinberg, Philip, and Berit Kristoffersen. 2017. "'The Ice Edge is Lost ... Nature Moved It': Mapping Ice as State Practice in the Canadian and Norwegian North." *Transactions of the Institute of British Geographers* 42, no. 4: 625–41.
Stengers, Isabelle. 2009. "History through the Middle: Between Macro and Mesopolitics: Interview with Isabelle Stengers by Brian Massumi and Erin Manning. *INFLeXions: A Journal for Research Creation* no. 3 (October). https://www.inflexions.org/n3_stengershtml.html.
– 2010. "Including Nonhumans in Political Theory: Opening Pandora's Box?" In *Political Matter: Technoscience, Democracy, and Public Life*, edited by B. Braun and S. Whatmore, 3–34. Minneapolis: University of Minnesota Press.
Strahler, Alan, and Arthur Strahler. 2005. *Physical Geography: Science and Systems of the Human Environment*, 3rd ed., Canadian version. Hoboken, NJ: John Wiley & Sons.
Strang, Veronica. 2004. *The Meaning of Water*. Oxford: Berg.
Stransbjerg, Jeppe. 2012: "Cartopolitics, Geopolitics and Boundaries in the Arctic." *Geopolitics* 17, no. 4: 818–42.
Strathern, Marilyn. 1992. *After Nature: English Kinship in the Late Twentieth Century*. Cambridge: Cambridge University Press.
– 1999. *Property, Substance and Effect: Anthropological Essays on Persons and Things*. London: Athlone Press.
– 2005. *Kinship, Law and the Unexpected: Relatives Are Always a Surprise*. Cambridge: Cambridge University Press.
Stunden Bower, Shannon. 2011. *Wet Prairie: People, Land, and Water in Agricultural Manitoba*. Vancouver: University of British Columbia Press.

Syvitski, J.P.M., and G. Robert Brakenridge. 2013. "Causation and Avoidance of Catastrophic Flooding along the Indus River, Pakistan." *GSA Today* 23, no. 1: 4–10.

Syvitski, J.P.M., and Albert Kettner. 2011. "Sediment Flux and the Anthropocene." 2011. *Philosophical Transactions of the Royal Society A* 369: 957–75.

Syvitski, J.P.M., Colin N. Waters, John Day, John D. Milliman, Colin Summerhayes, Will Steffen, Jan Zalasiewicz ... and Mark Williams. 2020. "Extraordinary Human Energy Consumptions and Resultant Geological Impacts Beginning around 1950 CE Initiated the Proposed Anthropocene Epoch." *Communications Earth and Environment* 1, no. 32: 1–13.

Szeto, Kit, Julian Brimelow, Peter Gybers, and Ronald Stewart. 2015. "The 2014 Extreme Flood on the Southeastern Canadian Prairies." In *Explaining Extreme Events of 2014 from a Climate Perspective*. Special supplement to the *Bulletin of the American Meteorological Society* 96, no. 12: S20–4.

Tansley, Arthur G. 1935. "The Use and Abuse of Vegetational Concepts and Terms." *Ecology* 16: 284–307.

Teller, James T., David W. Leverington, and Jason D. Mann. 2002. "Freshwater Outbursts to the Oceans from Glacial Lake Agassiz and Their Role in Climate Change during the Last Deglaciation." *Quaternary Science Reviews* 21: 879–87.

Teller, James T., and Zhirong Yang. 2015. "Mapping and Measuring Lake Agassiz Strandlines in North Dakota and Manitoba using LiDAR DEM data: Comparing Techniques, Revising Correlations, and Interpreting Anomalous Isostatic Rebound Gradients." *GSA Bulletin* 127, nos. 3–4: 608–20.

Thiessen, Kendall. 2010. "Stabilization of Natural Clay Riverbanks with Rockfill Columns: A Full-Scale Field Test and Numerical Verification." Unpublished doctoral thesis, Civil Engineering, Faculty of Graduate Studies, University of Manitoba.

Thomas, Andy, and Florence Paynter. 2010. "The Significance of Creating First Nation Traditional Names Map." *First Nations Perspectives* 3, no. 1: 48–64.

Thompson, Shirley. 2015. "Flooding of First Nations and Environmental Justice in Manitoba: Case Studies of the Impacts of the 2011 Flood and Hydro Development in Manitoba." *Manitoba Law Journal* 38 no. 2: 220–59.

Thompson, Shirley, Myrle Ballard, and Donna Martin. 2014. "Lake St. Martin First Nation Community Members' Experiences of Induced Displacement: 'We're like refugees'" *Refuge* 29, no. 2: 75–86.

Thomson, Tom. 1997. *Faces of the Flood: Manitoba's Courageous Battle against the Red River*. Photographs by Tom Thomson, text by Jake MacDonald and Shirley Sandrel. Toronto: Stoddart.

Toews, Miriam. 2022. "The Way She Closed the Door." *New Yorker*, 14 and 21 February: 18–26.

Tomiak, Julie, Tyler McCreary, David Hugill, Robert Henry, and Heather Dorries. 2019. "Introduction: Settler City Limits." In *Settler City Limits: Indigenous Resurgence and Colonial Violence in the Urban Prairie West*, edited by Heather Dorries, Robert Henry, David Hugill, Tyler McCreary, and Julie Tomiak, 1–21. Winnipeg: University of Manitoba Press.

Tsing, Anna. 2005. *Friction: An Ethnography of Global Connection*. Princeton, NJ: Princeton University Press.

Tsing, Anna, Heather Swanson, Elaine Gan, and Nils Bubandt, eds. 2017. *Arts of Living on a Damaged Planet*. Minneapolis: University of Minnesota Press.

Tsouvalis, Judith, Claire Waterton, and Ian J. Winfield. 2012. "Intra-actions on Loweswater, Cumbria: New Collectives, Blue-Green Algae, and the Visualization of Invisible Presences through Sound and Science." In *Visuality/Materiality: Images, Objects and Practices*, edited by Gillian Rose and Divya P. Tolia-Kelly, 109–32. New York: Routledge.

Tyrrell, Joseph Burr. 1890. *Geological and Natural History Survey of Canada, Annual Report*. New series, Vol. IV, for 1888–89. Montreal: William Foster Brown and Co.

Ullberg, Susann Baez. 2017. "Forgetting Flooding? Post-disaster Economy and Embedded Remembrance in Suburban Santa Fe, Argentina." *Nature and Culture* 12 (1): 27–45.

Upham, Warren. 1895. *The Glacial Lake Agassiz*. Washington, DC: US Geological Survey. https://play.google.com/books/reader?id=x0kzPVLsemkC&printsec=frontcover&output=reader&hl=en&pg=GBS.PA6.

Vaughn, Sarah E. 2019. "Inundated with Facts: Flooding and the Knowledge Economies of Climate Adaptation in Guyana." In *Unmasking the State: Politics, Society, and Economy in Guyana 1992–2015*, edited by Arif Bulkan and D. Alissa Trotz, 479–500. Kingston, Jamaica: Ian Randle Publishers.

– 2022. *Engineering Vulnerability: In Pursuit of Climate Adaptation*. Durham, NC: Duke University Press.

Vernadsky, Vladimir Ivanovich. [1938] 2001. "Problems of Biogeochemistry II: On the Fundamental Material-Energetic Distinction between Living and Nonliving Natural Bodies," translated by Jonathan Tennenbaum and Rachel Douglas. In *21st Century Science and Technology 2000–2001*, 21–39. https://21sci-tech.com/translations/ProblemsBiogeochemistry.pdf.

Vitz, Matthew. 2018. *A City on a Lake: Urban Political Ecology and the Growth of Mexico City*. Durham, NC: Duke University Press.

Vivian, Robert. 2001. *Des Glacières du Faucigny aux Glaciers du Mont Blanc*. Montmelian: La Fontaine de Siloé.

Warkentin, John. 1961. "Manitoba Settlement Patterns." In *Transactions Series III*, No. 16, edited by Douglas Kemp, 62–77. Winnipeg: Manitoba Historical Society. https://archive.org/details/trent_0116402015667/page/n1/mode/2up.

Weir, Thomas R., ed. 1983. *Atlas of Manitoba*. Winnipeg: Department of Natural Resources, Province of Manitoba.

Whatmore, Sarah J. 2013. "Earthly Powers and Affective Environments: An Ontological Politics of Flood Risk." *Theory, Culture & Society* 30, nos 7–8: 33–50.

White, Ed. 2012. "Flooded Manitoba Farmers Blame Western Neighbours." *The Western Producer*, 2 February. https://www.producer.com/news/flooded-manitoba-farmers-blame-western-neighbour.

White, Iain. 2010. *Water and the City: Risk, Resilience and Planning for a Sustainable Future*. London: Routledge.

Whitington, Jerome. 2018. *Anthropogenic Rivers: The Production of Uncertainty in Lao Hydropower*. Ithaca, NY: Cornell University Press.

Whyte, Kyle P. 2018. "Indigenous Science (Fiction) for the Anthropocene: Ancestral Dystopias and Fantasies of Climate Change Crises." *Environment and Planning E: Nature and Space* 1, no. 1–2: 224–42.

Williams, Mark, Jan Zalasiewicz, Neil Davies, Ilaria Mazzini, Jean-Philippe Goiran, and Stephanie C. Kane. 2014. "Humans as the Third Evolutionary Stage of Biosphere Engineering of Rivers." *Anthropocene* 7: 57–63. doi:10.1016/j.ancene.2015.03.003.

Williams, Raymond. 1977. *Marxism and Literature*. Oxford: Oxford University Press.

Winner, Langdon. 1977. *Autonomous Technology: Technics Out-of-Control as a Theme in Political Thought*. Cambridge, MA: MIT Press.

Withington, John. 2013. *Flood: Nature and Culture*. London: Reaktion.

Wood, Denis. 2010. *Rethinking the Power of Maps*. New York: Guildford Press.

Wynter, Sylvia. n.d. "Black Metamorphosis: New Natives in a New World." Unpublished manuscript.

Yazzie, Melanie K., and Cutcha Risling Baldy. 2018. "Introduction: Indigenous Peoples and the Politics of Water." *Decolonization: Indigeneity, Education & Society* 7, no. 1: 1–18.

Yogis, John A. 1998. *Canadian Law Dictionary*, 4th ed. Hauppauge: Barron's Educational Series.

Yusoff, Kathryn. 2013. Geologic Life: Prehistory, Climate, Futures in the Anthropocene. *Environment and Planning D: Society and Space* 31: 779–95.
– 2016. "Anthropogenesis: Origins and Endings in the Anthropocene." *Theory, Culture, Society* 33, no. 2: 3–28.
– 2019. *A Billion Black Anthropocenes or None*. Minneapolis: University of Minnesota Press.
Zalasiewicz, Jan, Mark Williams, Colin N. Waters, Anthony D. Barnosky, John Palmesino, Ann-Sofi Rönnskog, Matt Edgeworth ... and Alexander P. Wolfe. 2017. "Scale and Diversity of the Physical Technosphere: A Geological Perspective." *The Anthropocene Review* 4, no. 1: 9–22.
Zeiderman, Austin. 2016. *Endangered City: The Politics of Security and Risk in Bogotá*. Durham, NC: Duke University Press.
Zeller, Suzanna. 2000. "The Colonial World as Geological Metaphor: Strata(gems) of Empire in Victorian Canada." *Osiris* 15: 85–95.
– 2006. "Humboldt and the Habitability of Canada's Great Northwest." *The Geographical Review* 96, no. 3: 382–98.
Zilberg, Elana. 2017. "Bridging Divides at the Los Angeles River: Race, Nature and Infrastructure." Paper presented at the Association of American Anthropology, Washington, DC, 2 December.

Index

aesthetic, power: as action upon senses, 64, 225n17; backstage 94, 252n30; diatoms, 147; of diked landscape, 30; erratics, 241n4; of Lake Agassiz, 61; official discourse, 167; rivers, 13, 30; theatrical floods, 178; trilobites, 25, 28

Agassiz, Glacial Lake: as cultural stage, 72–5; in geographical imagination, 60, 243n25; life and death of, 5–7, 56–60, 64–7; and namesake's racism, 223–4n8; as research object, 45, 54; in science and engineering, 103–4, 151; in settler chronology, 27

agencies, riverine and lacustrine, entwined with cities, 1–2; of social actors/bodies, 116, 147–8

alliance of engineering and law, 29, 38, 189; and geoscience, 126, 134; and internal legal frontier/infrastructural edge, 138–40; and system balance logic, 128

Anthropocene: and Manitoban history, 27–9, 199–206; potential, 23–9, 229n41

aquatic nature: natural disaster as delusion, 165, 197; technical definitions of natural flooding/levels, 36, 60, 98–9, 110, 253n33

archive and Anthropocene dialogue (Appendix I), 199–206

art, 268–9n26; on Alpine ice edges, 243n29; and geological imagination, 170; as timespace materialization, 270n33; *Watermarks* by Mary Miss Studios, 251–2n27. *See also* earth art; erratics; films; literature; theatre

Assiniboine River. *See* River, Assiniboine

avulsion and Assiniboine: about, 181–94, 270–1n3; and engineering possibilities, 188, 191, 273n13; ethnography, 191–2; at The Forks, 182; undiscussed alternatives, 189

Bakhtin, Mikhail, M., 44, 158, 171, 174, 230–1n48, 265n4, 267n15, 268n23, 268n25, 270n33

balance: of currents/clay, 103–4; and extrapolation, 91; as good

life (Anishinaabe), 29; of gravity/buoyancy (isostatic rebound), 193; as ideal/hydrological measure, 29, 115–16, 127, 131; in lakes, 42, 144; in public debate, 195; and risk, 141; of uncertainty/action, 158, 170–1
Ballard, Myrle, 123–4, 129, 137, 259n15
Barad, Karen, 43, 142–3, 150, 262n1, 262n3, 267n15, 270n1
Barry, Andrew, 78, 90, 131, 136, 207, 210, 245n1, 255n17, 267n15, 279n1
being and becoming: in lawscapes and timescapes, 266n10, 267n15; river cities, 3–4, 20, 28, 181, 198, 221–2n1, 222n3, 228n34, 279n1
Bennett, Jane, 72, 125, 148, 221–2n1, 245n44, 262, 264n12
blended habitus: about, 42, 78, 156, 246n3; and avulsion's remix potential, 189; as default settler approach, 125; hybrid knowledge, interpretation, and action, 16–17, 27, 91, 93, 113. *See also* expert-inhabitants; focal points, as civic communication infrastructure; habitus; measurement
boundary-crossing and -making: animate/inanimate, 75, 148; and avulsion (law), 270–1n3; Canada-US border, 18, 227n28; in ethnographic stories and art, 237n88, 243n29; in fresh/saltwater diatom assemblages, 150. *See also* internal legal frontier
Bourdieu, Pierre, 189, 222n3, 228n34, 238n90, 246n3. *See also* habitus

breach, (un)controlled, in Hoop and Holler: in 2011, 135; in 2014, 135, 160–71; fears of, 164; in fiction, 172–3, 178; mitigation and compensation for, 164, 168–9
breakup (of river ice), as spring event and geological drama, 71–2, 84–5, 159; as turning point in personal history, 84. *See also* ice age end

calculations: artificial vs natural flood damage, 36, 59–60, 118, 125–9, 133–8, 165; that erase Indigenous practices, 125, 259n19; mass balance and probability, 184; railway as machine for, 61; Water Balance Methodology and out-of-whack lakes, 127–8
capitalism, logic and logistical power of, 61; as space of accumulation, 131; transnational, 68
chronotope: definition, 158, 265n4; of encounter, 171, 268n23, 178; of threshold, 172, 268n25
cities. *See* river cities
climate change/crises, 192–3; and colonialism, 29; fish migration, 193, 275n22; global shipping, 68; hydrogeology, 234–5n74; modeling assumptions and predictions, 102, 112, 189; as natural (dis)order, 83, 183; and resiliency, 272n10; river avulsion and engineering design, 185, 187, 189; summer rain-driven floods, 157, 159, 179, 265–6n7. *See also* ice age end; IPCC (International

Panel on Climate Change); water (in)security
collaboration: collaborative consent and UNDRIP, 29, 253–4n1; in a hybrid socio-hydrology project, 246n3; among Indigenous scholars and allies, 232n56; scientific, 43
colonialism. *See* climate change/crises; decolonization; engineering; flood control; Indigenous rights and activism; internal legal frontier; settler colonialism
communication, emergency. *See* focal points, as civic communication infrastructure; press conference on TV, flood emergency (2014)
conatus, 72, 221–2n1, 244n39
continental divide, Northern, 227n33; and cultural differences, 228n35
culture, as material social processes, 11
cuts, categorical: last cut of controlled breach, 164, 179; in law 126; as proposed start of Anthropocene (1950), 28. *See also* calculations; Wall Through Time; water/land separation
Cyclotella agassizensis: taxonomic power of and confusion with *C. quillensis*, 149–51

Da Cunha, Dilip, 19, 77, 122, 124, 186, 227n19
data collection/interpretation/modeling (geoscience): of ambiguity, complexity, probability, and spontaneity, 128, 184; of anomalies vs recursive patterns, 170; in intergenerational North American corpus, 45, 62–4, 67; isobases and interpolation debate, 61, 62–4, 69, 91. *See also* Upham, Warren
decolonization, 23–5; of flood control, 196–7, 235–6n76. *See also* Hudson Bay Company (HBC); TRC; UNDRIP (United Nations Declaration on the Rights of Indigenous People)
Delta, Assiniboine, as geological space of encounter, 58–9, 185–6
diatoms, 9–10, 43, 142–54; classificatory and signifying powers of, 149, 153; disturbance and transsexuality, 152; EU Water Framework Directive, 265n20; and microscope invention, 147–9. *See also Cyclotella agassizensis*
dikes: vulnerability and maintenance (Assiniboine), 13–14, 163–4. *See also* Lyndale Drive
disaster: and infrastructural tipping points, 198, 254–5n11; politics, 99–100, 126–7; slow violence, 38–9. *See also* floods, extreme events in modern Winnipeg; infrastructural outsiders; mimicry, in Portage Diversion design; suspense; *Winnipeg Free Press*
dispossession, as default process that sediments inequalities, 124–5; and engineering system design, 137–8; state strategies in hydrological zones, 237n83
Doctrine of Discovery, Indigenous critique of, 17–18, 29, 139, 196–7

earth art, 241n4; and LiDAR, 271–2n9; and Smithson's provisional theory, 187; 272n10

ecology: affective and relational, 147, 263–4n10; ecological democracy, 29; of floating and sinking, 142–55; invasive species and biodiversity loss post-flood, 119, 124

engineering: and climate change readiness, 170, 189, 197–8, 268n22; colonial legacies of, 38, 124–5, 137, 196; as ecological by default, 141; improvisation, 109–13, 197; and military, 161–9; modernity and monumentality, 26, 196, 198 231n49; river interlinkage and diversion, 43, 40, 43, 132, 236n82, 238–9. *See also* alliance of engineering and law; flood control; MIT; political engineering

Elden, Stuart, 23–4, 101, 106, 221–2n1, 227n28, 245n1, 253–4n1

erratics: about, 75; among Alpine shepherds and naturalist-geologists, 46–7; in art, 240n3, 241n4; as deep time clues, 41; language, naming, and signifying power of, 46, 74, 198, 239n2; at Universities of Winnipeg and Harvard, 72–5

ethnographic writing, experimental, 23, 70, 123; in blank spaces, 71–2; as ventriloquism 75, 193–4

ethnography. *See* ethnographic writing, experimental; juxtaposition; methods, ethnographic

evacuations, 32, 111–12, 122, 124

expert-inhabitants, 10; interviews of, 248n12; Richardson and the Z-Dike, 111–13. *See also* Rannie, W. (Bill); Thiessen, Kendall; Topping, Steve; Warkentin, Alf.

facts: deep/surface, knowing/feeling, measuring/extrapolating, reading/constituting, 20–3, 63, 186, 262–3n5; as symbolic infrastructure, 26

Fairford River Control Structure and Emergency Canal, 124, 136; and climate change 132–3; target of settler engineers (1890s), 132

feminist science studies, 262–3n5; 263–4n10. *See also* materiality and new materialism; more-than-human beings; STS

figure/ground reversals, of everyday life and flood extremes, 13; in ethnographic writing, 70

films: *Flooding Hope: The Lake St. Martin First Nation Story* (Myrle Ballard, Shirley Thompson, R. Klatt) 123–4; *Summerland* (Jessica Swale), 51; *When the Levees Broke* (Spike Lee), 124

flood bowl, formation, 55–60; complexity, 82. *See also* thrown-togetherness

flood control, geo-cultural definition, 3–4

flood control design and operations: lakes as storage basins, 116, 117; and property regimes, 102–9, 163–4; rules of, 98–100, 118; seasonal routines, 82–92, 104–6; south-to-north and west-to-east axes, 79–80. *See also* Number-Feet James flood control system (structures), 77–100: diagram of, 80; and roads and railroads, 109–10; supplementary earthworks (mounds, ring dikes, sandbags), 81, 163–4. *See also* dikes; Fairford River Control Structure and Emergency Canal; Floodway, Red River; Lake Manitoba, and 2011 flood; Lake St Martin First Nation; Portage Diversion

flood control, law and policy of, 36, 134; Master Agreement on Apportionment (1969), 247n8; operations, 98–100, 118. *See also* alliance of engineering and law; law; lawscape

floods, extreme events in modern Winnipeg: in 1950 (Great Flood), 30–3, 41; in 1997 (Flood of the Century), 31–3, 102, 109–14, 228–9n40; in 2011, 117–29, 188; in 2014, 156–71; as part of family histories, 31. *See also* archive and Anthropocene dialogue (Appendix I); climate change/crises

Floodway, Red River, 31, 33, 35, 159

focal points, as civic communication infrastructure, 78, 93, 195, 251n27. *See also* Number-Feet James

folkways as floodways, 82, 93, 168, 195, 246n3. *See also* blended habitus; focal points, as civic communication infrastructure

Forbes, Scott, 117–21, 258n6, 258n8, 260n28, 273n13

forecasting: and airborne gamma survey, 86–7; authoritative doubt in, 69; Flood Outlooks, 84–7; generational differences in, 89; and interpretive sifting, 84–6, 198; measurement and visualization, 60–4; Operational Forecasts, 87–92; overland flows, 191. *See also* suspense

Forks, The, 23–7, 229n4, 2232n57. *See also* Wall Through Time

geo-culture: about, 10–12, 21, 27–8, 34, 74, 78, 108, 192; in ambiguous geographies, 23, 119; and climate change, 83; Dutch/global, 177; embedded remembrance in, 93, 251n6; geographic confusion in, 20, 227n32; geoscientific imagination in, 5–7, 64, 89, 184–5, 191; in *longue durée* and planetary future, 25, 84, 156, 25; terminology in, 39–40, 78, 194, 275n23. *See also* blended habitus; flood control; stories; theatre; throwntogetherness

geological actors: about, 5, 10, 22: as diatoms, 148–53; and erratics, 46; human-water relationship in deep/ethnographic time, 45, 71; as Lake Agassiz, 54

geological imaginaries, 11, 76; geological estrangement in slave-trade and settler colonialism, 235–6n76; geontologies, 11; and geo-politics, 340f; mythic Anthropos, 25, 230n46; in science and engineering, 103–4, 234n73. *See also* art; erratics

geoscience: as channel jumping experiment, 184; convergence with Indigenous knowledge, 68;

North American origin story, 5–7; and the sublime, 50. *See also* ice age end

geovisualization: double artifact, 66–70; hyperreality, 69; and microscopy, 150, 264n17; of surface/subterranean ice/water, 86; technologies as theory-machines, 61; of technozones, 209 (AII). *See also* LiDAR and the Lidaresque; Outburst Quartet; Wall Through Time

ghosts: of Assiniboine, 181–2, 233n61; of Lake Agassiz, 192; of Red, 31

glaciers, 5–7, 76; as writers of bedrock inscriptions, 55. *See also* ice age end

governance, water, 194–8: and geoscience, 59, 259–60n24; and lake language and level, 115–16; as political engineering, 44, 158, 171; property regimes 102–9; and transparency and accountability performances, 128, 134, 167, 258n12. *See also* flood control

habitus: about, 21, 78, 222n3, 228n34, 246n3; turns of (1997, 2014), 83, 170

Haraway, Donna, 23, 43, 142, 244n36, 245n44, 262, 263–4n10

Holocene epoch: about, 5–7, 65; on-again-off-again bi-directional relationship of Lake Manitoba and Assiniboine River during, 129–31, 151. *See also* cuts, categorical

Hoop and Holler. *See* breach, (un)controlled

Hudson Bay Company (HBC): Department Store, colonial icon donated to Southern Chiefs Organization (2022), 73, 244n42; and imperial/settler history, 204–5, 226–7n27

human footprint, in Manitoban river systems, 28, 138–40, 205–6. *See also* archive and Anthropocene dialogue (Appendix I)

hydraulics, definition, 86; and citizenship, 235–6n76

hydro(geo)logy: in emergencies, 86, 158; hydrological cycle,19, 144–5, 227n29

hydrometric river network, 83; composite image of, 96; gauge uncertainties not clarified by more data, 95–7, 252–3n32; vulnerabilities and conflict, 97. *See also* satellite mediation

ice, and flood control (jams, dams, and frazil), 35, 87, 104–5, 134, 136; Arctic sea ice and global shipping, 68; chokepoint icebreaker performance (Amphibexes and Wolverines), 272n11. *See also* breakup (of river ice), as spring event and geological drama

ice age end: in geoscience 5–7, 45–62, 64–7; and ocean currents, 58, 65; spring reenactment of, 35, 71–2, 192. *See also* Agassiz, Glacial Lake; Outburst Quartet

identity, space-based, 35–6; of bedrock, drift, and till, 55–6; and Lake Agassiz, 45. *See also* infrastructural outsiders indicators: diatoms (in ecology and geoscience), 143, 146–7, 149, 153; lake level (system balance), 115–16; river focal point (moving flood crest), 98 92–3, 257n3; tree bark (flood history), 31

Indigenous rights and activism: in academia, 221n1, 235–6n76, 235–6n76, 261n32; and colonial engineering legacies, 38, 195; in Shoal Lake 40 and Winnipeg's aqueduct system, 238–9n92. *See also* film: *Flooding Hope: The Lake St. Martin First Nation Story* (Myrle Ballard, Shirley Thompson, R. Klatt); law; settler colonialism; TRC; Wa Ni Ska Tan

infrastructural outsiders: about, 35: farmers and Z-dike, 113; Hoop and Holler 163; and sacrifice, 38, 237n83; (un)recognition and water activism, 195, 235–6n76. *See also* breach, (un)controlled, in Hoop and Holler; Lake St Martin First Nation; Morris, town-specific risks on Highway; Twin Lakes Beach, post-flood disaster

infrastructure. *See* flood control; infrastructural outsiders; infrastructure and environment (aquatic interfaces)

infrastructure and environment (aquatic interfaces): anthromes/biomes, 138, 228n38, 260n30; ecological impacts, 258n8 279n1; envirotechnical systems, 108, 245n2; socio-techno-nature, 4, 107–8; and sustainable development, 34

internal legal frontier: about, 42–3, 260n30; and settler colonialism 138–40, 196, 238n91

interpretation, frameworks of: 1950/1997 as before/after markers of engineering mastery, 33, 83; in media about engineering, 139; repetition as cue, 67; techno-science 10–11

IPCC (International Panel on Climate Change), 234–5n74

justice, social and environmental: and alliance of engineering and law, 189; and concepts of balance, 127–9. *See also* Indigenous rights and activism; law; spatial justice

just one rain away, in Flood of the Century 1997, 33; in 2014, 168. *See also* structure of feeling

juxtaposition: of community memories/infrastructural crimes 123–4; in double artifact experiment, 66–70; of empirical knowledge/sphere of unintentional agencies, 192; in holistic analysis 10; of incompatible data sets/legitimate critique, 149–50, 264n14

knowledge production: as discovery and invention, 147–8; in geoscience, 64, 70, 184, 193; as global and universal, 228n38, 61, 184. *See also* facts; juxtaposition; shape of uncertainty

Kulchyski, Peter, 70–2

Lake Manitoba, and 2011 flood,

117–22. *See also* Twin Lakes Beach, post-flood disaster
Lake Winnipeg. *See* waterspouts
lakes: as ecologies of light, 145; as equilibrators and integrators, 144–5. *See also* balance
Lake St Martin First Nation: and 2011 flood, 122–5, 196–7; evacuation to the city, 122, 124, 136, 258–9n13; as the technozone's raw edge, 122
law: and assumption of uniformity, 39; Indian Act, Indigenous critique, 138, 253–4n1, 260n29; infrastructural regulation, 125–9; material nature of, 38; mixed effects of, 121. *See also* plural legal cultures
lawscape: about, 38; atmosphere, legal withdrawal from, 126, 140, 261n35; as river confluence (image), 40; yellow ribbons as post-disaster recovery icons, 117, 121
Lefebvre, Henri, 17, 38, 50, 116, 131
LiDAR and the Lidaresque, 44, 184–7, 189, 191, 193, 226n22, 271n7, 275n21
Lightfoot, Sheryl, 18, 138, 196
literature: Faulkner's *The Old Man*, 172–3; Toews's "Ass River," 84. *See also* theatre
logistical power: of entwined flood control and transport systems, 109–10; of ice age end, 46, 65; riverine, 60, 77–8; as socio-technical-ordering, 113, 197, 255n17; sustainability of, 100; of water and gravity, 132
Lyndale Drive: engineering and planting in restoration, 102–6; as primary dike/promenade, 12–13, 30–32, 105; river history, 30

MacMahon, Doug, 161, 165, 167, 171, 179, 211–20 (AIII), 255–6n18
Manitoba, Province of, as metaphor, 139–40. *See also* archive and Anthropocene dialogue (Appendix I); Manitoba Hydro; maps and mapping; MIT; territorialization
Manitoba Hydro, 131, 236–7n82, 245n2, 276n26
maps and mapping, 11–12, 19, 29, 61, 67; as hydropolitics, 19; of probable inundation, 161,169; sleight of hand in, 67; and von Humboldt, 50, 61, 226n27. *See also* scalar mismatches
Massey, Doreen, 21–3, 42, 75, 83, 89, 191–2, 271–2n9
materiality and new materialism: about, 142–3; biogeochemistry of living matter, 43, 148–9; and event of place, 191–2; geo-politics, 192, 234n73; material semiotics, 151, 245n44; thing-power, 72. *See also* law
measurement: elusiveness and conflict, 99; as inertia-production 129; and linguistic traditions, 97–8. *See also* calculations; forecasting; snow
meso: -politics, 259–6n24; -scale, 102; and systemic imbalance, 133
methods, ethnographic: discourse analysis 165–9; fieldwork as embodied experience 16, 17,

70–2; focus and detachment of, 122–3; and the (non-)relational and non-connected, 257n3, 266n10; as roaming, 46, 72. *See also* photography; juxtaposition; visual methods
metropolitan nature, 13; and creative urbanity, 113; and enclave ecology 229–30n44
mimicry, in Portage Diversion design, 131; of emotional flood disasters in fiction, 178; of nature's recursivity, 158
Ministry of Infrastructure and Transport. *See* MIT
MIT: experts' work, 85–6; and LiDAR's revelations, 187–9; managing caveats and fuzziness, 128–9, 133–8, 167–8. *See also* engineering; forecasting; expert-inhabitants
more-than-human beings: and bison iconography, 161; as ethnographic/geological actors, 43, 142, 146, 196, 263–4n10. *See also* ethnographic writing, experimental
Morris, town-specific risks on Highway, 100, 110

NAFTA: and specificities of space, 68, 244n37
names, politics of science: and erratics, 46–7, 239n1, 239n2. *See also* Agassiz, Glacier Lake
Napoleon, Val, 26, 253–4n1, 260n29
narrative. *See* literature; methods, ethonographic; press conference on TV, flood emergency (2014); stories; theatre

nature: made arbitrary by engineering, 132; and queer crossings (animal/plant, biotic/abiotic, sexual/asexual), 126, 148, 150, 152; and urban erratics, 74. *See also* aquatic nature; metropolitan nature
Norwood. *See* Lyndale Drive
Number-Feet James, 41, 92–7; as iconic measure, 98; (re)negotiation of, 195; as traditional communique, 93. *See also* flood control design and operations; focal points, as civic communication infrastructure

Outburst Quartet: as geophysical metaphor, 192, 198; as postglacial origin story 5–7, 64–7
overland flooding: and avulsion, 187–8; as contradictory effect of transport infrastructure, 110–11, 196; as technical foreshadowing, 114; in windstorm, 118

Philippopoulos-Mihalopoulos, Andreas, 38, 126, 138, 140, 188, 221n1, 266n10, 275n19, 280n4
photography: aerial, 62–4, 82; as integrator of reality/identity, 31; and photographs of in-situ memories, 8–9, 32, 109–10, 123; of uncanny landscapes of loss, 31–2; 67, 110, 123–4
pivots and turns: in approach to uncertainty, 179; into avulsion's plurilinear possibilities, 181; in engineering decision-making, 99–100, 162; in habitus, 83, 231n49; between lake balance

and level as framing concepts, 127; into microscopic worlds, 145

plural legal cultures, 26; vs commonality claims, 196; and law as process, 253–4n1

pluriverse, the, and plurilinear river-city meshworks, 185, 190, 274n18

political engineering, 34–37, 43, 97–100, 180; and engineering risk, 107–8; and a more inclusive future, 198; passivity in, 139–40; in state-of-emergency, 158; and technozone decision-making, 36, 127. *See also* press conference on TV, flood emergency (2014); space, politics of

Portage Diversion: bi-directional geological precedents for, 59; reliance and vulnerability, 118, 121, 258n8, 273n13; spillway scene, 7–11

power: of avulsion, 182; geology as operation of, 27; as geontopower, 27; of rivers 71–2 (Dene)

precautionary principle, and engineering, 190; unequal application of, 126

press conference on TV, flood emergency (2014), 156–71; as chronotope of encounter, 171, 268n23; as performance of logistical power, 157; temporal jurisdictions of speakers, 162–3

railway: as measurement infrastructure, 61; and multimodal global shipping, 68. *See also* Winnipeg, city of

Rannie, W. (Bill), 111, 114, 130, 182; Assiniboine Valley geomorphology tour, 185–6, 270n2; on city infrastructure tour 13, 30, 233n62, 233n64, 255n14

Red River of the North. *See* River, Red

restoration, forest, 25, 107–9, 229–30n44, 197. *See also* River, Assiniboine; Lyndale Drive

River, Assiniboine: agency, 159; climate communiques 10, 161, 197; 156–71; and Lake Agassiz, 130; Restoration Project, 229–30n44; trickster nature/summer flood (2014), 10, 156–71, 182–3. *See also* avulsion and Assiniboine; Delta, Assiniboine; dikes

River, Mackenzie (Deh Cho), 70–2

River, Mississippi: 1927 flood in, 172–3; 1950 flood in, 178. *See also* Outburst Quartet

River, Red: ethnography in Red River Valley, 11–19; Riverview, 106–9. *See also* Floodway, Red River; ghosts; Lyndale Drive

river cities: as body multiples, 186; geological precedents in urban planning, 108, 132, 274n16; as human collectives and geological actors, 3, 181; and seasons, 83, 249n14. *See also* being and becoming; Winnipeg, city of

riverhood: ghostly causality of, 270n1; impulsivity and compulsivity, 3, 8–9, 31, 71, 181, 198; as plurilinear meshwork, 189; and tentacularity, 67–8, 244n36

sandbags, super-sandbags, and geot-

ubes, 117, 119–20; as nomadic defence, 42, 112–13, 228–9n40; as supplement and symbol, 81–2
satellite mediation, 83, 87, 94–7, 112, 241n4
scalar mismatches: in emergency, 102, 109–13; in ethnography, 18–19; in map reading and flood forecasting, 16–17; in routine, 101–9, 105–9; as tactical sites for unsettling, 102
science and technology studies. *See* STS
settler colonialism, 38, 138–40; 238n91; and settler geology, 27, 182–3. *See also* geological imaginaries; internal legal frontier; Wall Through Time; water/land separation
shape of uncertainty, 178–80; and climate change, 33; and geoscience, 184; Lidaresque redefinition, 189; as a reverse analytic insight, 179, 192; and visualization of frames and anchors, 165–7
Simpson, Leanne Betasamosake, 23, 123–4, 228–9n40, 256n1
snow: snowmelt as flood driver, 84–5; as technical puzzle (measurement) 88, 90–1
space, politics of, 17, 21, 107–8: ethnographic visibility of, 106; of hazard and possibility, 82, 129; in lake regulation, 116; and meaning-making between points 227n28, 62–4; production of space, 36, 116; and sovereignty/globality tensions, 68–9; at The Forks, 77

spatial justice, 37–9; climate change action as social/environmental justice, 37, 190; and Lake St Martin, 138, 196–7; and technozone politics, 197, 276n26
spatial orientation: and empire, 228n35; beside frozen rivers, 14–15; at The Forks, 182–3. *See also* focal points, as civic communication infrastructure
speculation and science, 11, 63, 88, 191
sphere of unintended agencies: about, 3–4, 156, 237n86; and avulsion, 183; and diatoms, 9–10; and flood forecasting, 87; in geoscience and engineering, 38, 45, 193; and infrastructural inequalities, 34–9; and Lake Agassiz, 62; and shape of uncertainty, 179
stories: data-stories, 193; with flood event as affective metaphor and frame, 178; in flood genres, 3–4, 156; spatialization in, 93, 251n26
Strathern, Marilyn, 44, 125–6, 129, 147, 270n34
STS: and ethnography of infrastructure, 245n1; and problem of non-connection, 266n10
structure of feeling: about, 238n90; as decolonization, 238n90; as just one rain away 238n90; as lake shape and level, 116; in literature and theatre, 171–80
suspense: applying operational rules, 99; and engineering's reinvention, 197–8; flood disasters, 113, 168; in forecasting, 85–6; and technological revelation, 87; theatrical floods, 178

sustainability: and avulsion's challenge to Room for the River model, 190–1, 274n16; and riparian forest restoration, 274n17; and social justice, 34
Syvitski, J.P.M., 28, 144–5, 147, 199–200, 204, 229n41, 273n13

technical traditions, 6–7, 98–9, 193, 231n49, 254n6
technozone: about, 77–8, 90, 136, 196, 207–10 (AII); and lakes, 116; negative internalized impacts, 104–6; political possibilities in, 276n26; and technosphere, 264n12, 279n1; and urban-hinterland dependencies, 141. *See also* geovisualization; logistical power; political engineering
territorialization: and flood control, 42, 105–6; as mundane muddling, 108–9; and railway, 61; and *terra nullius* in Indigenous critical theory, 23–4, 139, 221n1, 261n32. *See also* internal legal frontier; technozone
territory as political technology, 101–2, 106, 108, 129, 140, 196–7, 253–4n1
theatre: Ian Ross's *The Gap*, 173–7; Tennessee William's *Kingdom of Earth*, 177–8
Thiessen, Kendall, 103–7
throwntogetherness, 21–3; and decision-making, 91; and erratics, 75; of flood bowl, 114; and sandbags, 42; and vulnerability in emergencies, 109, 113
timescapes: about, 267n15; in Anthropocene, 190; of flood damage (ecological, social, and structural) 118–19
timespace, 222–3n4; historical and geological intersections in, 22–3, 28, 41, 241, 186, 197; of river action in press conference discourse, 161–3; and spiraling time (Anishinaabe), 274–5n18
Topping, Steve, 20, 37, 112, 161, 167–8, 171, 211–20 (AIII), 260n26, 272n11
transdisciplinarity, 40, 43, 46, 143, 148, 264n12
TRC, 18, 139, 201, 232n57
Truth and Reconciliation Commission. *See* TRC
Twin Lakes Beach, post-flood disaster, 117–22, 258n9; recovery funds for permanent vs legal residents, 121; windstorm lake levels (31 May 2011), 118, 135–6

uncertainty: as inherent to knowledge practice, 270n34; negotiation of, 158; three types in Flood Outlooks, 87–8. *See also* shape of uncertainty
UNDRIP (United Nations Declaration on the Rights of Indigenous People), 18, 29, 138, 200–1, 226–7n27, 232n57, 253–4n1
Upham, Warren: empiric-poetic description of a prairie mirage, 51; enacting the flood bowl, 47; findings, 5–60; and horse and wagon geology fieldwork, 48–54, 186

Victorian culture, 47, 54, 242n19. *See also* Upham, Warren

visual methods: in ethnography, 44, 279n34, 165–9. See also juxtaposition; photography; press conference on TV, flood emergency (2014)

Wall Through Time, 23–7, 230–1n48, 232n57
Wa Ni Ska Tan, 237n84, 261–2n36, 276n26
Warkentin, Alf., 20, 89–90, 112, 260n26
water (in)security: about, 70, 114; and avulsion, 183; and climate and engineering, 83; and confidence in government, 168; and decision-making fragmentation, 107–8, 249n23; senses, sensibilities, and dispositions, 26, 41
water/land separation: and dispossession of Indigenous people, 28, 138–40; and nature concept, 122, 124; un-settling of, 198. See also cuts, categorical
waterspouts: supernatural empirics of, 153–4
WEE (wind-effect-eliminated water levels), 125–6, 136; proposed reconsideration of, 196
wildlands, category obscuring urban infrastructural links, 260n30
wind: and lake ecology, 151–4; and rain, 112; and waves, 42, 117–22, 125–6. See also WEE (wind-effect-eliminated water levels)
Winnipeg, city of: and avulsion as potentially disruptive/creative force, 182–3, 188–9; CentrePort, 67–70; Exchange District, 242n18; geo-cultural achievements of, 34; model flood control system, pros and cons, 34–9; Naturalist Service Office, 103; railway and economic boom (1881–1911), 54; setting of Ian Ross's play *The Gap*, 173–7; survey map (1874–5), 29–30
Winnipeg Free Press: letters of Scott Forbes, 118, 258n6, 258n9; news images of Indigenous flood victims, 39, 122; as source and keeper of flood disaster memories, 82, 90
WSC (Water Survey Canada): and USGS cooperation, 83, 88. See also hydrometric river network

Yusoff, Kathryn, 11, 27, 41, 230n47